Spiritual Design

Enrich Your Spiritual Practice with Lessons from Behavioral Science

Stephen Wendel

Northeast Press

Copyright © 2019 by Stephen Wendel

All rights reserved. No part of this publication may be reproduced, distributed, or transmitted in any form or by any means, including photocopying, recording, or other electronic or mechanical methods, without the prior written permission of the publisher, except in the case of brief quotations embodied in critical reviews and certain other noncommercial uses permitted by copyright law.

Northeast Press
Oak Park, Illinois 60302

Cover Design by Erik M. Peterson
Edited by Elizabeth Zack and Tim Dedopulos
An Artisanal Publication

Printed in the United States of America
First Edition: September 2019

Publisher's Cataloging-in-Publication data

Names: Wendel, Stephen
Title: Spiritual Design / by Stephen Wendel.
Description: Oak Park, IL: Northeast, 2019.
Identifiers: LCCN 2018915146 | ISBN 9781733531207 (softcover)
Subjects: LCSH: Spiritual Formation | Spiritual Life | Economics--Psychological aspects | BISAC: RELIGION / Christian Living / Spiritual Growth.
Classification: DDC 248.4—dc23
LC record available at https://lccn.loc.gov/2018915146

ISBN: 1-7335312-0-3
ISBN-13: 978-1-7335312-0-7

For Mark

CONTENTS

	Preface	i
1	A Behavioral Understanding of the Mind	1
2	Designing for Spiritual Practice	25
3	Taking Action in our Spiritual Lives	35
4	Devoting our Time and Attention	47
5	The Community Around Us	57
6	Finding the Urgency to Take Action Now	67
7	Sustaining a Practice	81
8	Avoiding Bad Habits and Other Challenges	97
9	The Power, and Limits, of Design	111
10	Closing Thoughts	129
	Bibliography	135

PREFACE

As a behavioral scientist, I find remarkable parallels between the challenges of living a spiritual life and behavioral research on how people struggle to turn intentions into action. The book is about those parallels, how we can identify obstacles to our spiritual practice, and hopefully, with grace, overcome them. It's not about my personal journey, though I will share it along the way. It's also not about what you should believe or which church or religious institution you should attend. That's up to you. Instead, it's about how each of us designs our spiritual environment — and how we make it easier, or harder, for ourselves to engage in spiritual practice.

Before we dive into spiritual design itself, let me explain a bit about my own background, and what I'm trying (and not trying) to accomplish here.

Behavioral Science on the Job

For the last decade I've conducted research as an applied behavioral scientist. Behavioral science, if you haven't come across it, is an interdisciplinary study of how people behave in their daily lives, especially where those actions don't seem to be wise or in their long term-interest. Behavioral scientists have a particular understanding of how people make decisions (especially bad ones), a set of techniques to improve decision-making and behavior, and a commitment to rigorously testing these techniques in the field. The first chapter of the book provides a much more detailed look at behavioral science, with a range of examples. For now though, we'll leave it at that.

I started on this path at a small personal finance company that developed an online finance app to help people save and pay off debts. We quickly learned that (almost) everyone already knew what to do: spend less and save more. The problem was actually doing it. And that's where we devoted our time and energy: developing ways to help people do what they already knew they should do.

I later joined a much larger investment research and software company, where now I lead a team of researchers studying financial behavior. We run experiments in-house, and draw from the broader literature in behavioral science, to better understand the problems that people face in everything from saving, to investing, to working with financial advisers. I've had the good fortune to run hundreds of experiments over the years, and studied thousands more conducted by others in the field; these studies provide many of the insights offered in this book.

I've also had the joy of writing two books on applied behavioral science, one on applications to product development, and one specifically on how the benefits we each receive at work can be better designed and communicated.[1] Along the way (in 2012), I started a non-profit organization called the Action Design Network. Its intent is to bring together and learn from a broader community of behavioral practitioners. The group holds regular events on how to use behavioral science in software for social good. Thanks to our group — which now hosts events in ten cities around the country for over fifteen thousand people — I've been very lucky to learn from some of the brightest minds in the field. We've held conferences with academic researchers and practitioners from Harvard to Stanford, and from Instagram to Uber. It's been an amazing organization to work with.

---

[1] See Wendel (2013) and Wendel (2014)

As my professional path proceeded, I started to see increasing parallels between two parts of my life: my growing faith, and my research on human behavior. I saw that both share a deep understanding of our human imperfections and how our well-intentioned plans can go awry. In both communities, there's a belief that people can do better, given the right environment and circumstances. Naturally, there are a great many differences — each community has its unique stories and lessons to share.

This book is a form of translation, from one community to another. In it, I attempt to bring the tools and research lessons developed in behavioral science into a spiritual context. It's not a guidebook, and it's not a set of instructions: each person's situation and spiritual path is too unique for that. Hopefully, however, this book can offer some hints and ideas to help readers live more fulfilling and meaningful spiritual lives.

I believe that *lessons from behavioral science can help us close the gap between what we want to do and what we actually do, between our spiritual intentions and our actions.*

Closing the Intention-Action Gap

In my research, I'm particularly passionate about helping people take action. I focus on what's known as the **intention-action gap**: the often-considerable gap that exists between what people *want* to do (for example, to exercise more, pray regularly, or learn a new language), and what they *actually* do. Depending on the particular behavior, that gap can be truly massive! One illustrative study showed how the gap between the number of people who intended to save more for the future, and those who actually did, was a whopping 90%.[2]

Thankfully, behavioral science has a lot to say about this problem and has numerous tools that can help with bridging the gap. These tools

---

[2] Only 10% of the people who committed to do it actually followed through. See Choi et al. (2002)

are particularly interesting for two reasons. First, the solutions are often non-obvious. For example, I often hear that someone didn't follow through on their intention to act because they "just weren't motivated enough." In reality, motivation often isn't what's actually lacking. Instead there a myriad of other non-obvious issues that get in the way, like personal insecurities, seemingly trivial barriers, or how the task is presented. Second, the solutions are empowering — in the non-cheesy, literal way. They empower people to take actions with which they otherwise struggle. They aren't coercive, and they aren't even persuasive; they are about *helping people do what they already want to do*.

Armed with the behavioral understanding of why we stumble, and research on how to overcome these problems, we can use that knowledge to do better. For example, if we're afraid that we'll forget some of the groceries we wish to pick up, what do we do? We write down a list or set one up with Alexa or an app. We have our spouse send us a text message to remind us. These are simple ways, as researchers call it, to bring something to "top of mind" so that we are more likely to remember it and act on it. Behavioral researchers have studied such simple interventions as sending text messages to help people remember to put aside money for savings, and found them to be effective.[3] More generally though, if we forget to do something repeatedly, often the simplest solution is to build a reminder into our environment, such a reoccurring calendar item, or even a note on the fridge. When it comes to our everyday lives, we intuitively know this.

When it comes to our spiritual lives, however, we treat the gap between our intentions and our actions differently. When we fail to follow through on a religious commitment we've made, we sometimes take it as a sign of being spiritually insincere, insufficiently motivated, or simply being a failure. In reality, perhaps we just have a faulty memory (like everyone else in the world) that affects our ability to follow through on all of our commitments, spiritual or otherwise. If we allowed ourselves some "environmental assistance" — the spiritual equivalent of a note on the fridge — we could focus more on our

---

[3] Karlan et al. (2010)

spiritual journey, and not on why we are such terrible people who don't follow through on the things that we said we would do. Behavioral science can help us work around our limitations and better translate our intentions into action.

Unfortunately, there's no simple guide on how to do that. Despite a wealth of research across other facets of our lives, behavioral science hasn't studied how people can live more spiritually meaningful lives. The topic is perhaps too sensitive, too "unscientific."[4] If nothing else, it would be ethically troubling for researchers to pick a religious practice out of a hat, experiment on people to see if they would adopt it, and measure how it affects them. Thankfully, that isn't the only option. We can take the behavioral literature as a set of insights and tools, to see how they might apply to and enrich our spiritual practice. If something doesn't resonate, leave it be. If it does, give it a try, and see where it takes you.

In my own life, that's what I've tried to do. Over the years, I've sincerely wanted to spend time in prayer, know God and live a more spiritual life. But I've struggled, with my intentions never quite materializing into real change in my daily discipline and practice. In the end, I discovered only God could help me change. Along the way, though, I helped prepare the path by focusing my attention on God, and making it slightly easier for myself to listen and to learn from Him. In so doing, I found how lessons from behavioral science could make sense of, and help, each major milestone in my religious life. And now I sincerely hope that behavioral lessons also can help you interpret and pursue your path to a more meaningful spiritual life.

Our actions make faith and faithfulness more or less likely to blossom. We can find creative and sincere ways to pursue our commitments, whether those commitments are to explore what faith means to us, to return to God after a long absence, or to deepen our daily practice. A behavioral approach can help us overcome some of our challenges, and give us the strength to move forward, through the process of spiritual design.

---

[4] Though interestingly, it's now accepted and common practice in psychology to study the drivers of personal happiness.

God and Our Religious Traditions

For most of my life, I've been a Quaker, and because of my Quaker faith, I think in terms of God, the Holy Spirit, the *Inward Light* (a common Quaker expression for either the in-dwelling Christ or the Holy Spirit), and, to a lesser extent, the Son, Jesus Christ. I recognize that other people use other words for God, and focus on God's manifestations differently. For Evangelical Christians, there is a strong focus on Jesus Christ. Among the Eastern Orthodox, there is often a focus on the Holy Spirit. In other faiths, like Islam and Judaism, those communities use other terms for God. People who have decided against organized religion sometimes avoid such terms altogether, and use terms like The Spirit or The Divine. Even within Quakerism, there's a great diversity of terms and of theology.

While my background is Quaker, in all likelihood yours isn't. For better or worse, there just aren't that many Quakers in the world, and I'm not going to trying to convince you of Quakerism or talk much about Quaker history and theology. Instead, my goal is to share what I've seen through the lens of behavioral science, to help you on your own path of spiritual deepening and spiritual discipline. This book is about living according to your faith, regardless of the details of your specific religious beliefs.

Throughout this work I'll draw on lessons from the Scriptures, from Quaker writings, and from across the Christian Church. My wife and I are now part of a wonderful Lutheran Church, and I'll share experiences from that world too. In addition, many years ago I tried practicing Zen Buddhism, and I will draw upon a few lessons from that community. From my limited knowledge, there are very similar lessons in many other religions — from Judaism to Hinduism. I'll speak primarily to Christianity, though, simply because that is what I know.

In addition, I'll use terms like "God" and the "Holy Spirit," but not because you must have a Quaker (or even a broader Christian) understanding of the Divine. Rather, I'll use them because of my own religious background; that's simply how I know God myself. Similarly, I will refer to God as "He" because that's how I understand God, and not because I think you need to use that pronoun. I mention this because I know that referring to God as "He" is foreign to some people — yet that's just what I know and how I was taught.

Please do not let my expression impose on yours, any more than I would want yours to impose on mine. We each have our own way of describing God. We must each express our understanding as authentically as we can.

What to Expect

Throughout the book, I'll offer examples of how behavioral science can shed light on spiritual practice. In particular, I'll discuss how we can design our environments to help us spend more time and attention on spiritual matters and to put our faith into action. As I mentioned earlier, there's no manual on how this works, nor is this book meant to be one exactly. Instead, it offers *lessons from behavioral research translated into the language and context of spiritual practice.*

To provide a context for these lessons, Chapter 1 offers a practical introduction to behavioral science, starting with a behavioralist's understanding of how people make decisions in their daily lives and cleverly use mental shortcuts to work around their limitations. These clever shortcuts, or *heuristics,* mean that what we feel, what we believe, and how we act in daily life are shaped to a surprising degree by our environment. It also means that changes in that environment can change — but not determine — our actions and decisions.

In Chapter 2, I introduce the concept of spiritual design: how people can intentionally design their physical environments, social milieu and daily schedules to support faith and faithfulness. The techniques of spiritual design vary depending on the type of action one is trying to take, from one-time actions, to building habits, to avoiding problematic behaviors. These different goals define the structure of the next few chapters.

Chapter 3 starts a discussion on how to "start" doing something: from praying more regularly to responding to a personal calling. It introduces the EAST framework, an acronym that researchers sometimes use for the major tools of behavior change: make it *Easier* to act, grab *Attention*, leverage *Social* influence, and make the action *Timely* or otherwise urgent. Chapter 3 then delves into the 'E' in EAST (making it Easier), and provides practical examinations of techniques you can try in your own life. Chapters 4, 5 and 6 cover the other three dimensions: Attention, Social Influence and Timeliness. We review nineteen techniques in all, each of which have been field tested by researchers in other domains, and which we can experiment with and learn from in our own spiritual lives.

Chapter 7 addresses the particular challenges of building up a regular routine, or to move from "starting" something to "continuing" something. We look at how habits are formed in the brain, and how we can keep ourselves on track with feedback loops and thoughtfully writing our self-narrative. We also examine what happens when we falter, and how to benefit from behavioral "fresh starts" and reapply ourselves to the task.

Chapter 8 tackles the other side of the behavioral coin: breaking bad habits and avoiding moments of weakness we come to regret later. Bad habits are particularly pernicious because we often can't overcome them with sheer willpower. The reality is, they aren't necessarily defects of character, but rather result from vital cognitive mechanisms gone awry. When we understand how bad habits are triggered, we can more wisely work around them.

Chapter 9 explores a range of the limitations of spiritual design, from the vast differences we each have in our spiritual lives — and how a particular technique may not only not resonate with us, but may backfire — to the particular challenges of deepening our faith. We look at how we should be very cautious applying spiritual design to anyone beyond ourselves, including our own children. We also explore a case study of the power and limits of spiritual design from the unprogrammed Quaker tradition that I come from. Under different names, spiritual design has played a major role in the spiritual growth and practice of Quakers, and has shaped the evolution of that community — with great benefits, and some real and unintended consequences.

After this analysis, the last chapter provides a few closing thoughts for my readers. It discusses the role of academic research, and how lay people can develop and expand upon these lessons as we learn more about the power of spiritual design. There are many open areas where practical, thoughtful research could help spiritual communities grow and thrive. Similarly, individuals can and should find the path that best helps them deepen their spirituality, hopefully sharing those lessons with the broader community.

In this book, I sincerely hope you will find inspiration for your own path. This writing has helped deepen my own spiritual life. May it help you deepen or rekindle yours as well.

ACKNOWLEDGEMENTS

Thank you to the early readers who gave feedback on the germ of an idea, and later on the outline and draft: from my Earlham School of Religion classmates and my teacher, Stephen Angell, to scattered friends and fellow travelers in the behavioral science community including Nicolae Naumof, Daniel Crosby, Dan Egan, Paul Davies, Kristin Aldred Cheek, and Karla Paxton.

For my friends Paul Rosenfeld, Justin Thorp, and Micah Bales, who inspired me as I first started on my own spiritual path: thanks for taking me along with you.

And for those who pushed me to complete the book, and brave my fears about publishing it — from Ryan Murphy to Rob Pinkerton: thank you for the careful nudges.

This book is possible because of my wife, Alexia, and her love.

1
A BEHAVIORAL UNDERSTANDING OF THE MIND

Although humans are not irrational, they often need help to make more accurate judgements and better decisions.
— Daniel Kahneman, Thinking, Fast and Slow

Do you have a friend who has a gym membership, but just doesn't go very often? Does that person enjoy giving money to a gym they don't use? Of course not. They really intended to go when they first signed up. They still know that it's what they need, and that they'd probably enjoy it. Despite all of their past failures, they keep hoping and believing that they will get it together and go regularly. So, they keep paying — and keep failing to go. *Good intentions, and the sincere desire to do something, aren't enough.*

There's an interdisciplinary field of research called behavioral science that studies why this occurs, along with many other seemingly irrational or strange behaviors. I'm a behavioral scientist, and I research situations like these. Behavioral scientists call not-going-to-the-gym-but-still-paying-for-it an example of an **intention-action gap**. The intention to act is there, but people don't follow through and act on it. It's not that people are insincere or lack motivation; the gap happens because of how people's minds are wired. Behavioral scientists have developed a detailed understanding of how the mind works, and what drives day-to-day behavior, including the intention-action gap.

Behavioral scientists have also developed many techniques to overcome such quirks of the mind. One such technique is known as **temptation bundling**, in which we only allow ourselves to do something we really want to do (like listen to a trashy audiobook) while we're doing the thing that we really should do and intend to do (like going to the gym). Or we put our gym clothes near the door as a prompt to go work out. Behavioral researchers test ideas like this with formal experiments on randomly selected groups of people to rigorously determine what works and what doesn't. The goal is to help people overcome their behavioral obstacles.

Do you see a parallel between going to the gym and spiritual practice? If not, think about this: do you have a friend who sincerely became a member of a church or another spiritual community, but just doesn't seem to go very often? Most people know someone like this. Does that friend enjoy being "a part of" a community with which they rarely interact? Probably not. In all likelihood, they really intend to go and seek the fellowship and support of others. The person knows that it's what they need spiritually, and that they'll probably enjoy it. And, despite all the evidence to the contrary, they keep on hoping and believing that they will get it together and go regularly.

This parallel occurs because *we face the gap between our intentions and actions in many areas of our lives, no matter the context.* We're just imperfect human beings, and that's true across the board. The same patterns of behavior — the seemingly irrational, foolish and flawed things we do — occur again and again.

While this fact can be depressing, there is also hope: lessons from behavioral science about how to overcome or work around these failings can help, both in everyday activities like going to the gym, and in our spiritual lives. We can learn to follow through on our yearning to meditate, to read Scripture, to attend church, or to express gratitude — rather than hoping that the human imperfections that hinder us from doing so will simply disappear.

What Behavioral Science Has to Offer

Behavioral science is an interdisciplinary field that in particular combines cognitive and social psychology with economics. Over the last decade, there has been a tremendous growth of research in this field and also of best-selling books that share research lessons, such as: Dan Ariely's *Predictably Irrational*, Richard Thaler and Cass Sunstein's *Nudge*, and Daniel Kahneman's *Thinking, Fast and Slow*.[5] Thaler and Kahneman have each won the Nobel Prize, in large part for their work in behavioral science.[6]

One of the most active areas of research in behavioral science is how our environment affects our choices and behavior, and how a change in that environment can then change those choices and behaviors. Environments can be thoughtfully and carefully designed: to help us become more aware of our choices, to shape our decisions for good or for ill, and to spur us to take action once we've made a choice. We call that process **choice architecture**, or **behavioral design**.

Behavioral scientists have studied a wide range of behaviors, from saving for retirement[7] to exercising. The earlier example about temptation bundling is from one such study about exercise.[8]

---

[5] Ariely (2008), Thaler and Sunstein (2008), Kahneman (2011)

[6] Behavioral science is not 'behaviorism', by the way. Behaviorism was an older tradition in psychology, associated with BF Skinner, which focused solely on external behavior and denied the importance of internal cognitive mechanisms. Behavioral science, sometimes shortened to 'behavioralism' draws from cognitive psychology, and focuses heavily on people's (imperfect) cognitive mechanisms. Two similar names, but very different perspectives.

[7] Thaler and Benartzi (2004)

[8] Milkman et al. (2013)

Researchers have found many ingenious ways to help people take action when they would otherwise procrastinate or struggle to follow through.

...And What It Doesn't

Unfortunately, there doesn't appear to be any behavioral research specifically on deepening one's spiritual practice.[9] While many (non-behavioral) researchers have studied religion, they haven't done so in a way that we might apply. A range of researchers have looked at how one's religiosity relates to, and perhaps causes, behaviors such as cooperation, charity, and trust.[10] That work is interesting; it's just not useful as a practical guide to spiritual practice. Other researchers have looked at how our minds might be genetically wired for religion, and how our brains might process religious concepts.[11] Theologians and scientists have also examined the act of conversion.[12] None of these are particularly helpful for the lay practitioner.

Not only isn't there a "behavioral science guide to spirituality," this book does not try to offer one. That research simply hasn't been conducted. Instead, this book is a two-way translation between the language of spirituality and faith on the one side, and behavioral science on the other. As I've followed my own spiritual path, I've found ideas from both domains to be immensely useful. I offer these translations in the hope that others might find them rewarding as well.

What are some of those lessons? Let's start with the seemingly simple, but profound: *our environment affects our ability to live meaningful spiritual lives.* This is especially true for the daily habits we create, the people with

---

[9] The most relevant work is likely Ken Evers-Hood's (2016) analysis of how pastors can learn from behavioral science to lead their churches with compassion and wisdom.

[10] Tan (2012)

[11] E.g., Jones (2015)

[12] E.g., Peace (1999), Baer et al. (2014)

whom we interact, and the reminders and distractions we face in our environment. By surrounding ourselves with people of faith, for example, it can become easier for us to deepen our own. By intentionally disrupting our routines, we can become more aware of our actions and create an opening for new spiritual experiences. Our daily rituals of sleep and meals can make it easier — or harder — to resist the status quo in our lives and try a new direction. By reflecting on the actions we've already taken that show faith, we can change our self-narrative, and bring ourselves further along our path. We'll talk about these and other approaches throughout the book.

One important area we won't talk about is one's inward experience of God. Whether one does or doesn't experience direct communion with God is a very different thing than our environment affecting our spiritual (and other types of) behavior. As you'll see, making decisions and acting according to our spiritual intentions have parallels in behavioral science; communion with God does not, and so we won't discuss personal experiences of the Divine. Instead, we'll discuss behavioral science as it relates to intentionally designing the structure of one's time, physical environments and social milieu to support faith and faithfulness. And that process starts by examining our simple human limitations.

A Short Story About Our Limitations

Imagine a guy who's just made a New Year's resolution to eat healthier. We'll call him Jim. Jim starts off well, cooking for himself, and fighting off the urge to binge on junk food. But, at some point, he walks into the kitchen, and sees a bag of chips. He picks it up and starts munching on the chips without a second thought. Once he realizes what he's doing, he curses at his slip-up. Unfortunately for him, eating is mostly habitual — people don't actually think about it, and often they just do what they've always done automatically. In disgust with himself, Jim throws out the bag of chips, and creates a list of healthy things to buy at the store.

The next day, Jim goes to the grocery store… but he forgets his shopping list at home. Luckily, he remembers some of the items and he knows what healthy food looks like, right? He walks over to the cereal aisle, and he's overwhelmed. Normally he just buys the same thing every time; today he reads what's on the boxes. "Maybe this one is healthy? Fiber is good, right?" Jim starts looking for words like "healthy" and "nutritious." He picks up a cereal box with the wording "healthy flakes," along with some other things he needs, and heads for the register. Yet his simple shortcut for buying healthy food has backfired: the cereal he selected is packed full of sugar (to hide the fact that the "healthy" ingredients taste terrible). But Jim doesn't know that. He's done his sincere best.

Jim's well-meaning attempt to set a new, healthier, path for himself goes awry — not because he's insincere, lazy, or foolish, but simply because of how his mind is wired. Jim's human.

We can use Jim's example to illustrate some key lessons from behavioral science about how people make decisions and behave:

- People don't carefully and rationally think through much of their daily behavior. Instead, we're of **two minds**: a slow, deliberative and effortful process that works through the costs and benefits of a choice, and a fast, effortless process that relies on a mix of habits, intuition, and emotion. For example, eating is primarily habitual;[13] as Jim found out when he tried to "eat healthy."

- Even when we're thinking carefully and rationally, **our minds are simply limited**. We have limited memory and attention; we procrastinate; we aren't great about doing math or complex problems in our heads. Jim, like every other normal person, forgets things. In this case, unfortunately, he forgot his shopping list.

- We're not dumb, though. In the face of our limitations, **our minds have developed clever shortcuts** which help us make

---

[13] E.g., Riet et al. (2011)

pretty good decisions in most circumstances.[14] For example, a simple shortcut we use to figure out what to do in a new situation is this: look at what others are doing, and mimic it. Such shortcuts are often quite helpful, but they can fail us — as they did for Jim. Jim's mind tried to use the shortcut of looking for the words "healthy" and "nutritious" on cereal boxes. He hoped that seeing these words was a sign that the food was actually healthy and nutritious, and not just a marketing ploy.

- Because of our minds' shortcuts and limitations, the **details of how a choice is presented matter immensely**. For example, our "eat chips" habit is more likely to be triggered if a bag of chips is within sight —when we walk into the kitchen. Marketing ploys, like "healthy flakes," work because a few words on a box can change our choices too.[15] The company selling the cereal knew that and took advantage of Jim's perfectly reasonable shortcut.

- And finally, when we understand this process, **we can recognize the mind's frequent and predictable errors, and we can overcome them**. And that — for Jim and for us too — is the point of this book.

These lessons are part of a behavioral understanding of the mind: our limitations, why people make seemingly foolish mistakes, and how we're often not thinking carefully and calmly. *This understanding can help us look at our spiritual practice in a different light.* Armed with these lessons, we can better identify why we fail to follow through on our spiritual commitments, and we can build the spiritual practice we desire — because the obstacles we face often aren't due to a lack of motivation or insincerity. Rather, like Jim, we often falter simply because of how our minds are wired. And thankfully that doesn't mean we're stuck. We can wisely work around our limitations.

---

[14] Kahneman (2011)

[15] E.g., Alba (2011)

Making Things Better for Jim (and All of Us)

This *understanding* is one of the three major parts of behavioral science. The second part is a set of *tools* that can help people fix those mistakes, and the third part is a commitment to assessing those tools using rigorous *experiments*. Each of these parts is fascinating, but the integrated whole has proven to be a tremendously powerful approach to understanding, and overcoming, behavioral obstacles that people face in their daily lives.

The tools that behavioral scientists have developed to help people are truly impressive. Some of them are exotic and fascinating. For example, there are **Odysseus Contracts** in which people pay money to remove their freedom of choice in the future, because they know they will be tempted to do foolish things.[16] The name comes from Odysseus (Ulysses to the Romans) in Homer's epic poem *The Odyssey*. Odysseus had his shipmates tie him to the mast of his ship, so he couldn't respond to the alluring song of the deadly Sirens. Before he was tempted, he knew that he might succumb to a moment of weakness in the future, and he prepared accordingly. In the same way, we can intentionally "tie ourselves to the mast" when we want to resist future temptations that might cause us to stray from our spiritual path.

Other tools that behavioralists use are so simple that we tend to forget how powerful they are. For example, one of the easiest ways to help people overcome problems of attention and memory is to just **use a reminder:** there's much thoughtful research literature about the humble reminder.[17] Similarly, there's research on how we can become overwhelmed by too many options, and how to **simplify choices** to make it easier to decide and so act on those choices, rather than procrastinating. Overall, behavioral scientists study the impact and

---

[16] Ashraf et al. (2006)

[17] E.g., Karlan et al. (2010)

circumstances under which a range of techniques help people make better decisions and act on them.

Behavioral scientists are forever testing the actual impact of this impressive array of tools on the lives of real people — and that's the third aspect of behavioral science. We tend to favor the most rigorous tool out there: the randomized control trial. In a **randomized control trial**, some people are randomized assigned to receive a potentially effective tool, while others are randomly assigned to not to. The random assignment process means that between the two groups — those with the tool, and those without — there are no *aggregate* differences between the groups at the start of the experiment. The two groups have the same age distribution, the same proportion of women and of men, the same level of interest in sports; the same *everything*. Thus, any aggregate differences that occur after the test must be because of the tool itself, and nothing else.

Randomized control trials allow the researcher to say with confidence whether the tool actually *causes* people to improve or not.[18] They are the "gold standard" for figuring out whether something genuinely has an impact on behavior and outcomes — that is, whether or not it "works."

Putting these three pieces together, the day in a life of a behavioral scientist goes something like this: find a problem worth solving, try to understand what behavioral obstacle (if any) is causing it, come up with a potential solution that works around that obstacle, and then test to see whether it works in practice. That's a recipe for learning about people and solving practical problems that behavioral scientists have applied over and over, to great success. At the very moment you are reading this, there are behavioral scientists following something akin to this path at universities, national governments, and private companies around the globe.

Over the last decade, the field of behavioral science has virtually exploded in size and scope, from researchers studying how to prevent

---

[18] E.g., Gerber and Green (2012)

domestic violence in New South Wales, Australia, to people studying why rainy days make Airbnb clients less happy with their rentals (and what to do about it).[19] In large part, the rapid growth of research and understanding comes from the fact that a wide range of people have been able to apply this recipe to provide new insights and answers to their specific behavioral challenges across the globe and across diverse domains. Our goal here is lay the foundation for similar insights into the behavioral challenges that hinder one's spiritual practice.

A Closer Look

Now that you have a sense of what behavioral science is, let's dive deeper into the behavioral understanding of the mind. As mentioned above, we can draw out five major lessons:

- We're of two minds, intuitive and deliberative;
- Even when we are at our best, our minds are simply limited;
- We use shortcuts that can help us, but sometimes they go awry;
- Because of our minds' limitations and shortcuts, the details of our environment can matter immensely; and
- We can cleverly design our environment to avoid problems with our mental limitations and shortcuts gone awry.

Let's start with the first piece.

Being of Two Minds

What does it mean to "be of two minds?" While the analogy is imperfect, you can think about our minds as having two means of making decisions and of acting. When we're faced with a problem or a decision to make, both of them start working on it, at the same time.

---

[19] Behavioural Insights Team (2017); Pearson (2014)

One of them, the reactive system, is lightning-fast: it makes snap judgments and gives you an intuitive answer to what you should do. The other is more careful and deliberative but takes a lot more time (here, "a lot" can signify additional seconds or hours!). We call these two systems — the reactive and the deliberative systems — System 1 and System 2.

When you're driving a car and someone stops in front of you, you immediately *react* and slam your foot on the brake — that's your System 1. It takes in a small amount of information — seeing something blocking your way — and triggers a response: *Brake!!* It doesn't wait to think through the type of car that's ahead, it doesn't think about what the insurance bill might cost if there's a collision, or whether you're wearing your seatbelt. Your System 1 commands your foot to hit the brake. The rapid response is immensely valuable; It can save our lives. It also gets us into trouble — like, perhaps, when it leads us to get angry, in a split second, and say something rude to someone who cut in front of us in line.

Our reactive system doesn't simply control sudden movements and emotions, like braking or getting angry. It is also the source of many of our intuitive judgments, such as when we "just know" or "feel" that someone is a good person. Or when we get a bad feeling in the pit of our stomach about someone. Such commonly experienced reactions are *nonconscious*, but not in a Freudian way; rather we just don't consciously think about and cause them. Nevertheless, they guide our behavior.[20] We're nicer and more open to the first person we assessed as "good," and more wary with the second, because of our gut feelings. The logic behind these intuitive judgments isn't directly accessible to our minds. So, we might react to a situation based on an intuitive feeling and not know *why* we felt or acted that way. That is especially true with another form of System 1 behavior: habit.

We use the word habit in daily life in many different ways to describe everything from the "things we do pretty often" to "a chemical or psychological addiction." Here, we'll use a specific definition of a habit

---

[20] E.g., Damasio et al. (1996)

Spiritual Design

as studied in the behavioral literature, which is: *actions that we've repeated enough times, in the same context, that our minds automatically trigger them when facing that context.*

When we act on habit, our conscious minds might be thinking about something completely and utterly different in that moment. For example, driving a car is habitual for many people. This habit allows us to seamlessly respond to cars around us, automatically making the thousands of minute muscle movements that control the steering wheel, brake, acceleration, etc. to keep us on the road with minimal attention. While driving, we are often thinking about something else entirely. In this way, we really are "of two minds" which operate simultaneously and relatively independently: the conscious thinking part, and our nonconscious responses, including habits.

The way habits work is by basically outsourcing control over them to the environment.[21] Cues in the environment tell our minds when to execute the habit. Habits can be immensely useful and valuable for that reason: they radically decrease the amount of attention and scarce cognitive resources we need to devote to the task. For example, when we see a red light ahead of us while we're driving, the mind recognizes that cue and executes the push-down-on-the-brakes routine with lightning speed and without thinking over the costs and benefits of doing so.

When we do something again and again, in the same environment, we slowly build up a habit. In the case of driving a car, or riding a bicycle, initially we expend a lot of energy to ensure that we don't crash. We do the best we can. Over time, though, we learn how to be more effective drivers or bicycle riders, and build up our habits. Eventually, we just don't think about maneuvering the car or bike, and yet we're far better at it than when we first had to think about it. The same thing happens with playing a sport, for example. Eventually our play becomes more effortless, nonconscious and skilled.

We can choose to build good habits over time. For example, we can intentionally build a habit of giving our spouse a kiss when he or she

---

[21] Wood and Neal (2007)

comes home from work. Each evening when we see the car or hear the footsteps, we can go up to the door and greet him or her lovingly. The more often we do this, the less we will have to think about it, and the less cognitive "effort" will be required. Eventually when we see the car and hear the footsteps at the right time of day, we'll automatically stand and go to the door. Building habits in this way can be a real benefit when we are trying to do something useful or good.

Unfortunately, in the same ways that our minds build good habits, they also build bad ones. Bad habits can be extraordinarily difficult to overcome, and for the very reason good habits are so useful: we don't consciously think about them! Our bad habits trigger automatically in the context in which we learned them. That context might be driving home from work and seeing a certain bar we've been to too often, which triggers the habit of stopping in for a drink (or two or three). For some people, the context may be the seeming insubordination of an employee or even a child, which triggers an angry response, even if that insubordination was caused by the person simply not understanding what we were asking.

In hindsight we may berate ourselves for being foolish, and for making these mistakes. But in reality, acting on bad habits in this manner isn't really a "mistake": it's a *response*. The real mistake comes from not understanding how to avoid or overcome these bad habits — which thankfully behavioral science has quite a lot to say about.

Now, let's talk about System 2, our deliberative thinking. System 2 is what we normally mean when we refer to "thinking": it's the type of thinking that we're aware of in our conscious minds. It involves *working through a series of steps in our heads*, thinking over a potential action or decision and perhaps weighing the costs and benefits of each option. It's slow. It takes effort. And, while it's more careful and can come to a more appropriate choice than a rapid-fire reaction or habit, it's not perfect. That's because, even when we're thinking carefully about something, our minds have many, many limitations.

Our Minds Are Simply Limited

Remember Jim? The second problem he faced was that he forgot his grocery list. Who hasn't forgotten something at some point in their lives? Heck, who hasn't forgotten something in the last hour, or the last five minutes? I certainly have…

Forgetfulness is one of our many human frailties. Personally, the older I get, the longer that list seems to grow. There are sadly many ways in which our minds are limited and lead us to making choices which aren't the best. We fail to live out our intentions for a few broad reasons, which include the limitations of:

- Our attention
- Our cognitive capacity
- Our memories

And, these limitations string together. In terms of our *attention*, there are nearly an infinite number of things we could be paying attention to at any moment. We could be paying attention to the sound of our own heartbeat, the person who is trying to speak to us, the interesting conversation someone else is having near us, or the report that's overdue and we need to complete. Unfortunately, researchers have shown again and again that we can really only pay attention to one thing at a time. Despite all of the discussion in the popular media about multitasking, multitasking is a myth.[22] Certainly we can *switch* our attention back and forth; we can move from focusing on one thing to focusing on another — and we can do so again and again and again. But the reality is, switching focus frequently is costly: it slows us down, and it makes it harder for us to think clearly.

Given that we can only focus on one thing at a time, and that there are so many things that we could focus on (many of them urgent and interesting), it's no wonder that sometimes we aren't thinking about the promises we've made to ourselves — even if that promise was a

---

[22] Hamilton (2008)

spiritual one to meditate or pray at bedtime. Or to momentarily pause to experience the wonder and awe of God's creation all around us before we hurriedly go to work.

Similarly, our *cognitive capacity* is limited: we simply can't hold many unrelated ideas or pieces of information in our minds at the same time. You may have heard the famous story about why phone numbers in the U.S. are seven digits plus an area code: because researchers found that we could hold seven unrelated numbers in our heads at the same time, plus or minus two.[23] And, of course, there are so many other ways in which our cognitive capacity is limited. For one, we have a particularly difficult time dealing with probabilities and uncertain events, and with realistically predicting the likelihood of something happening in the future. We tend to over-predict rare but vivid and widely reported events, like shark attacks, terrorist attacks and lightning strikes.[24]

In addition, we can become overwhelmed or paralyzed when faced with a wide range of options, even as we consciously seek out more choices and options. Researchers call this the "paradox of choice":[25] our conscious minds believe that having more choices is almost always better, but when it actually comes time to make a decision and we're faced with our limited cognitive capacity and the difficulty of the choice ahead of us, we balk.

Lastly, when it comes to our memories, they simply aren't *perfect* and nothing is going to change that. For most of us, having a "not perfect" memory is a significant understatement: we are quite forgetful… Our memories usually aren't crystal-clear videos, but rather are a set of crib notes from which we reconstruct mental videos and pictures. We remember events that occur frequently (like eating breakfast) in a stylized format, losing the details of the individual occurrences and remembering instead a composite of that repeated experience.

---

[23] Miller (1956)

[24] E.g., Manis et al. (1993)

[25] See Schwartz (2004), Iyengar (2010)

Additionally, in some circumstances, we remember the peak and the end of an extended experience, not a true record of its duration or intensity.[26]

What do all of these cognitive limitations mean? They are important for two reasons in particular for our spiritual lives. First, these cognitive limitations mean that *we sometimes just don't make the best choices:* we don't take the actions that are in our best interest. We might want to pray regularly, we might want to give to charity, but we don't. It's not that we are bad people; it's that we are, simply, people. We get distracted, we forget things, we get overwhelmed. We shouldn't interpret a few bad choices as a sign that we, or someone else, is fundamentally a bad person; instead, it's just that our simple human frailties may be at work.

Second, *we often don't make conscious, thoughtful decisions about what actions to take in our daily lives.* It would simply be too overwhelming to think through the probability of each possible outcome for a single action, let alone to handle the full complexity of a choice among the nearly infinite things that we might do in any given moment. Our minds protect their limited cognitive capacity by trying to "economize" on thinking. Habits are one way in which our minds try to economize, but they're not nearly the only way. When we're thinking deliberatively and, especially, when we're thinking intuitively, we use a variety of shortcuts to economize on thinking: to reach a decision fast.

Our Minds Use Clever Shortcuts

Our minds' myriad shortcuts help us sort through the range of options we face us on a day-to-day basis, and make rapid, reasonable decisions about what to do. Here are a few of the commonly used shortcuts:

- **Descriptive norms.** If we aren't sure what to do, we look at what other people are doing and try to do the same (a.k.a. *descriptive norms*).[27] This is one of our most basic shortcuts.

---

[26] Kahneman (2000)

[27] Gerber and Rogers (2009)

For example: "People from here quote from the Bible, so it's okay for me to do that as well."

- **Availability heuristic.** When things are particularly easy to remember, we believe that are more likely to occur.[28] For example, if I'd recently heard in the news about a student being punished for praying in school, I'd naturally think that that is much more serious or common than it actually is.
- **Halo effect.** If we have a good assessment about someone (or something) overall, we sometimes judge other characteristics of the person (or thing) too positively — as if they have a "halo" of skill and quality.[29] For example, if we like someone personally, we might overestimate their skill at dancing, even if we knew nothing about their dancing ability.
- **Confirmation bias.** We tend to seek out, notice and remember information already in line with our existing thinking.[30] For example, if someone has a strong political view, they may notice and remember news stories that support that view and forget those that don't. In a sense, this tendency allows our mind to focus on what appears to be relevant in a sea of overwhelming information. (It's also quite troubling, since we ignore new information that might help us gain a truer picture of the world or try new things.)
- **IKEA effect.** When we invest time and energy in something — even if our contribution is objectively quite small — we tend to value the resulting item or outcome much more.[31] For example: after we've assembled IKEA furniture, we often value it more than similar furniture someone else assembled (even if it's of higher quality) — our sweat equity doesn't matter in terms of market value, but it does to us.

---

[28] Tversky and Kahneman (1973)
[29] Nisbett and Wilson (1977)
[30] Watson (1960)
[31] Norton et al. (2012)

- **Mere exposure effect.** The more we're exposed to something, like an idea or an object, the more we tend to like it (all else being equal).[32] For example, advertisers rely on this principle when they buy ads to show you an image of a brand again and again — just by seeing the ad, people can come to like the brand more (again, all else being equal).

There are over a hundred of these shortcuts (called **heuristics** in the research literature) or other tendencies of the mind (called **biases**) that researchers have identified. Unfortunately, these shortcuts can also lead us astray as we try to make good choices in our lives. For example, if you're a spiritual person living in a place where people don't speak about faith, descriptive norms apply a subtle (or not-so-subtle) pressure to avoid doing so yourself. Or a homeless person might look and smell dirty, and the (negative) halo effect could lead others to think negatively about them; they might see the person as less honest and less smart than they really are. While I've now mentioned some negative outcomes to our shortcuts and biases, it's important to understand that, at their root, our shortcuts are clever ways to handle the limited resources that our minds have.

Another reality is this: *we can't avoid using shortcuts completely*. Rather, by understanding how our shortcuts are triggered by the details of our environment, we can learn to avoid some of the negative outcomes that result from their use.

Remember, the Details of our Environment Matter Immensely

And finally, we turn to the last big lesson: *the importance of our environment on our behavior*. What we do is shaped by our environment in obvious ways, like when the architecture of a building focuses our attention and activity toward a central courtyard. It's also shaped in non-obvious ways: by the people who we talk with and listen to (our *social*

---

[32] Zajonc (1968)

environment), by what we see and interact with (our *physical environment*), and the habits and responses we've learned over time (our *mental environment*). These non-obvious effects can show themselves even in slight changes in wording of a question. Let's take a look at one example.

Suppose there's an outbreak of a rare disease, which is expected to kill 600 people. You're in charge of crafting the government's response.

You have two options:
1. Option A will result in 200 people saved.
2. Option B will result in a one-third probability that 600 people will be saved, and a two-thirds probability nobody will be saved.

Which option would you choose?

Now suppose there's another outbreak, of a different disease, which is also expected to kill 600 people.

You have two options:
1. Option C will result in the certain death of 400 people.
2. Option D will result in the one-third probability that nobody
dies and two-thirds probability everyone dies.

Which option would you choose now?

Presented with these options, people generally prefer Option A in the first situation, and Option D in the second. In Tversky and Kahneman's famous study using these situations,[33] 72% of people choose A (versus 28% for B), but only 22% choose C (versus 78% for D). Which, as you've probably caught on, doesn't make much sense, since for both A and C, four hundred people face certain death, and two hundred will be saved. Logically, if someone prefers A, that person should also choose C. But that isn't what happens, on average.

Many researchers believe there is such a stark difference in people's choices for these two mathematically equivalent options (A and C) *because of how the choices are framed.* One is framed as a *gain* of two hundred lives, and the other is framed as a *loss* of four hundred lives.[34] The text of C leads us to focus on the loss of four hundred lives (instead of the simultaneous gain of two hundred) while the text of A leads us to focus on the gain of two hundred lives (instead of the loss of four hundred). And people tend to avoid uncertain or risky options (B and D) when there is a positive, or "gain," frame (A versus B), and seek risks when faced with a negative, or "loss," frame (C versus D).

That's, well, *odd*. It shows how relatively minor changes in wording can lead to radically different choices. However, it is especially odd since this isn't something that people would explain themselves. If they were faced with both sets of choices, they wouldn't say, "Well, I recognize that A and C have exactly the same outcomes, but I just intuitively don't like thinking about a loss, even when I know it's a trick of the wording." Instead, the person might simply say: "knowing that I can save people is important (A) and I really don't like the thought of knowingly letting people go to certain death (C)."

---

[33] Tversky and Kahneman (1981)

[34] Many researchers accept this explanation, but not all. As often happens in science, there is a divergence of opinion on why framing effects like this occur. An alternative view is that people make a highly simplified analyses of the options and the two different options have two different simplified answers. See Kühberger and Tanner (2010) for one such perspective.

Stepping back from this particular example, sometimes we *explain* our behavior after it's happened, yet we don't know the real reasons ourselves. We are, as social psychology Tim Wilson nicely puts it, "strangers to ourselves."[35] Numerous experiments demonstrate how our minds make up stories to explain our behavior.[36]

This lack of self-knowledge also extends to what we'll do in the future. We're bad at forecasting the level of emotion we'll feel in future situations, and generally bad at forecasting our own behavior in the future.[37] For example, people can significantly overestimate the impact of future negative events on their emotions, such as a divorce or medical problem.[38] We're not only affected by the details of our environment, but we don't often recognize that our environment has affected us in the past, and so we don't consider the influence when we're thinking about what we'll do in the future.

How Context & Environment Affects Our Moral Behavior

Researchers find that the environment we're in shapes not only everyday behavior, but also moral behavior. A long history of research shows *how factors in our environment shape whether we act responsibly or not*.[39] For example:

- When people hear the cries of someone having an epileptic seizure in another room, the more people that hear, the less likely it is that anyone responds.[40]

---

[35] Wilson (2002)

[36] See Nisbett and Wilson (1977b) for an early summary

[37] Wilson and Gilbert (2003) (emotion); Wilson and LaFleur (1995) (behavior)

[38] See Wilson and Gilbert (2003) for this and other examples.

[39] See Appiah (2008) for a summary.

[40] Latané and Darley (1970)

- People are more likely to cheat on a test when they can't be caught, when they see others cheat, and when they can rationalize it as helping someone else.[41]
- People are more likely to help someone who gives a meaningless reason for asking for help, versus giving no reason. In one famous study, for example, people were more likely to allow someone to cut in line and use a copy machine when that person said they 'have to make copies' (which is obvious) versus when that person simply asked to use the copy machine.[42]

My personal favorite example of how factors in the environment shape our behaviors is the story of the seminary students.[43] In that study, researchers had seminary students do an activity, at the end of which they had to go to another building. They did not know that the travel to the other building was, in fact, the key part of the study itself.

The researchers varied the degree of urgency with which the students we asked to move, and varied the activity the students undertook before traveling to the other building. In one version of the pre-travel activity, the students prepared to discuss seminary jobs; in another, they got ready to discuss the story of the Good Samaritan. In addition, they were asked to go to the other building with one of three different levels of urgency. In each case, however, the seminary studies passed by a man slumped in an alleyway. He moaned and coughed, and the researchers had observers record whether the seminary students stopped and helped the individual.

It turned out the level of urgency mattered: the more urgently the student was supposed to reach the other building, the less likely they stopped to assist. The pre-travel activity (thinking about the Good

---

[41] Ariely (2013), which also provides a nice summary of how people deceive themselves in daily life: what exacerbates that (ambiguity, cheating to help others, seeing others cheat), and what minimizes it (clear feedback on what dishonesty is, supervision / surveillance).

[42] Langer et al. (1978)

[43] Darley and Batson (1973)

Samaritan story) did not. The ineffectiveness of thinking about the Good Samaritan story makes the research more dramatic and interesting, but the truly important finding was how simply being asked to hurry changed the behavior of presumably moral and helpful people. Specifically:

- In the least-urgent situation, 63% of the people helped the man slumped in an alley.
- In the medium-urgency situation, 45% stopped and helped.
- In the highest-urgency situation, 10% did.

A temporary, and in the grand scheme of things, largely unimportant detail (whether the person was asked to hurry or not) had a massive effect on the person's behavior. To put it bluntly: such things *shouldn't* matter — not if we're good and thoughtful people, right? Yet they do. And as much as we might condemn the students in this famous study, I'm sure we can all remember similar times in our lives as well, when we had something on our minds and didn't take the opportunity to help someone in need.

The research on moral behavior, and how it's shaped by our environment, ranges from the comical to the truly troubling, for it raises serious spiritual concerns. How can people be "moral" in one situation, but "immoral" in another situation that's only slightly different? Yet perhaps this is well in line with the Bible's understanding of how we are all imperfect people....

It also raises questions about what it means to be a good or moral person. Perhaps to be a good, we need to do more than be good in the moment. We should plan for good, and make it a structural part of our lives. As philosopher Kwame Appiah argued, compassionate people can use this research to create a 'perceptual correction' on their own experiences, to better understand the role of context and situation on their moral behavior. Or as Appiah's colleague and fellow philosopher

Gilbert Harman advised, we should:[44]

> Put less emphasis on moral education…and more emphasis on trying to arrange social institutions so that human beings are not placed in situations in which they will act badly.

Moral behavior obviously has many more drivers than simply our situation. But this research helps us to turn our attention to the (partial) role that situation and environment play in our lives. And it helps us see how we can design that environment to follow moral teachings, or simply to reinforce a spiritual practice in our lives, such as reading Scripture. That's what we'll turn to next.

---

[44] Appiah comment and Harman quote are from Appiah (2008)

2
DESIGNING FOR SPIRITUAL PRACTICE

The architecture always wins.
  — Traditional Christian saying, about how the architecture of a church shapes events within it[45]

In a Bare Room...

Imagine that you walk up to an old wooden porch and into a church that's just one large central room. As you enter, dozens of people are sitting in rows facing each other with their eyes closed and heads bowed. Everyone is quiet and the windows are open. You take a seat along the wall. You peek around, and see men and women, from young kids to grandparents, dressed in no particular fashion. There's nothing special in the room: no podium, no choir or band, not even any posters or large crosses on the wall. Just people sitting and waiting in the silence.

---

[45] My thanks to Pastor Dave Lyle for sharing this maxim.

Spiritual Design

As you settle in, and try to still yourself, sounds you can only hear in the quiet emerge. You hear the lovely rustling of autumn wind in the dry oak and poplar leaves outside. The old woman sitting next to you is slowly and effortfully breathing. The service starts.

As you sit there, you try to calm your mind alongside your body. You may recite a prayer in your head or focus on a verse of Scripture. You breathe gently and feel the wonderful gift of life. Your mind certainly wanders — to what you need to do after service, to something rude someone said to you — but you try to bring yourself back. This time is a time devoted to God. To contemplation of Him and to listening for God's word.

In time, someone a few rows over stands up. She waits a few moments, then starts. Maybe it's a spiritual song. Maybe it's a short recollection of something she's seen. Maybe it's a statement about our world, and what it means. But either way, it's something she feels the Spirit is moving her to say — something that should be said. And when what should be said is said, she sits down. Over time, perhaps other people stand, speak, and sit again.

At an appointed time, the contemplation ends. Each person reemerges from the silence, and church business is discussed.

That is what it's like to attend service at an unprogrammed Quaker Meeting. I first experienced it in high school, and have continued, throughout my life, to savor that unique setting for seeking God. This particular example came from Sandy Spring, a beautiful, simple, and deeply spiritual Meeting in Maryland.

There's a great diversity of Quakers, even of Quakers in the tradition of unprogrammed Meetings. But the design and experience at Sandy Spring and other many unprogrammed meetings is both distinctive and intentional: *there's nothing on the walls because what matters is God.* Early Quakers wanted to remove earthly distractions from their contemplation and communion with God. Quakers face each other, instead of a central podium, because early members of the church saw this as an expression of how we are equal before God. We each strive to follow Jesus's path from our imperfect selves towards his perfection.

And there's no preacher because the Holy Spirit is available to us all, if we simply listen to the still small voice within.

The Depths of Ritual

You're in a pew facing a raised altar. The woodwork is amazing: delicately carved oak fluted columns, robed saints looking to one another and to heaven. On your left comes filtered blue, red and purple light through windows of apostles and their converts.

As everyone gathers, a vast crowd fills the pews. Kids jostle. Old friends smile. And in time, all turn towards the entrance. A calm and careful teenager carries a massive cross into the room, followed by pastors and church leaders. In the balcony, a choir starts singing with the ease and beauty that comes from years of practice.

As the lead pastor reaches the front, he addresses the congregation: You are forgiven. You are loved. And through Jesus Christ, you can learn to walk in his ways.

Three times, you hear the Word of the Lord, and you offer thanks to God. The pastor helps illuminate the Word, to bring it closer to your life and your world. You recite the creed, give thanks for all that you have been given, and are welcomed into the mystery of the Eucharist.

When you were a child, perhaps this was a time to color in pictures of Jesus or furtively look around at other kids. Maybe you absolutely hated having to sit still and be quiet as people droned on up front.

But now, as an adult, it's different. The elements of the liturgy are the same every week, and now you know them, and they are part of you. Over the course of the hour, they simply flow out from you, to join with the somber tones of countless others, unified into a powerful whole. From your baptism to today, you are part of a community of believers. You reinforce one another in the faith, looking out for one another, and growing in Christ. Each Sunday, you enter this special place, one that looks like nothing else in your daily life. The ritual, the community, the manifest presence of God and His Church: they all feed you and help you turn to God throughout the intervening week.

For many people, this example will be the more familiar one. This experience is one my family and I share each Sunday with hundreds of members of the Lutheran Church near our house. Our church takes the liturgy seriously: not as a burden or an empty tradition, but as a means of making the sacred real and meaningful in our lives. The songs we sing, the rhythm of the church calendar, the shared expressions of our faith are all there to allow us to better focus our hearts and minds on God.

For me it's hard to imagine church experiences that are more different than the Lutheran and Quaker ones. But both are powerful and meaningful in their own way, and both are *intentional*. The emotions they evoke, and the spiritual focus they foster, don't occur by accident. Each is the expression of a thoughtful community of believers, looking to sustain and grow their faith.

Each is an expression of spiritual design.

Spiritual Design

Spiritual design is the act of intentionally creating an environment that supports spiritual practice or growth. We can see it at work in collective worship: from the bare simplicity of my Quaker background, to the rich liturgical traditions of the Lutheran Church, to the rock bands and riotous praise of non-denominational Protestantism. The simplicity, liturgy and rock music are different approaches to supporting the faithful.

When we step into a house of worship, especially one that's new and different to us, the collective choices and patterns of that community are often strikingly obvious. And while we may disagree with — or even viscerally reject — how a particular house of worship and its traditions are designed, we intuitively understand that somewhere, sometime, people thought about creating that environment. Those choices helped create the community that worships within, and often were centered around the specific needs of that community.

Spiritual design in collective worship is not an abstract thing, but rather a practical way to help a particular group of people connect with God. It's the same God, whether we are in a bare room or an ornate church. However, across spiritual communities, the needs and understandings of how to best feed our souls differ. Two very different designs can both support the growth of communities, or even the same community at different points in time. Hopefully the design of each serves the needs of those people at that time.

Communal spiritual design is fascinating — but not something most of us can affect. However, *spiritual design is also something we can apply in our daily lives as well*. We can create an environment at home, in our social communities, in our daily commute, and at work, that can remind us that there's more to this life than just what we can see and touch. That there's meaning, compassion and love to be found through living our faith in our families and in community with one another.

In this book we'll talk specifically about spiritual design at an individual level — since many of you probably aren't pastors thinking about how to redesign a church service, and because that's where we can all start, in small and big ways, to improve our spiritual lives. You'll learn how each of us can apply spiritual design.

A Simple Example

Have you ever placed a Bible on your bedside table? Not just after you'd been reading it, but because you wanted to read it? I have, and I know many others have as well. It's a minor action that changes a small detail of our daily environment. But it can help us follow through on our desire to read the Scripture, especially when we might otherwise get distracted or forget. The reason it can help us follow through isn't happenstance. There's a logic and rationale behind it.

The Bible on the bedside helps by serving as a *reminder* to read the Scriptures, and by *removing friction*, or obstacles to action. Since the Bible is already there, you don't need to go looking for it. These are behavioral lessons — about the gap between our intentions and actions, and what to do about it — applied to spiritual life.

And there are many other examples of how changes to our environment can help us follow through on a particular spiritual discipline, or deepen our focus on spiritual matters, such as:

- **Create a clear, distraction-free space to facilitate meditation** — something you can see in early Quaker meeting houses and practice, and many Zen Buddhist centers, for example.
- **Focus on a particular person or group of people** you want to aspire to be like in your spiritual practice, instead of implicitly comparing yourself to society's shallow examples of success.
- **Carry a symbol of your devotion and practice** — as a cross around your neck, or prayer beads on your wrist — to remind yourself of that part of your life and bring it to the fore (what's known as *self-signaling*).[46]

You've probably come across such recommendations in other contexts, and you may have tried some in your spiritual practice. One goal of this book to show you how many such techniques share a common foundation: *they are based on how our minds make decisions and take action*. And, to allow you to move beyond commonplace tips, another goal of this book is to help you apply this knowledge more intentionally to your own spiritual life and overcome the obstacles you face. You'll come to appreciate how you can apply lessons from behavioral science to intentionally design your environment to support your spiritual practice and growth.

The areas we'll focus on include:

1) Our physical environment: the things we see around us.
2) Our social environment: the people we talk with.

---

[46] Bodner and Prelec (2003)

3) Our mental environment: what we focus on and think about.

In each case, spiritual design can help us become slightly better at turning faith into action, and at weaving faith into our daily lives and thoughts. I write "slightly better" because this process doesn't assume we have complete control over our lives; no human does. We can do our part, and have faith that God will do His. Spiritual design also doesn't mean ignoring the power of God's grace: instead, it means recognizing that we have the free will to choose our path and we must *use* that free will, employing the gifts we have been given.

Preparing the Way

The idea of changing one's environment to support spiritual practice may strike you as a contradiction. For many people, faith and communion with God is something we can't will into existence: it is a divine gift. Personally, I agree. Our relationship with God is not something we can dictate. To believe that we can control it is a delusion; God must make it possible. Sometimes we're shaken to our core to pay attention, and sometimes we can just barely hear the voice within. Either way, God makes our relationship possible through grace.

Nevertheless, I believe we have a part to play. In particular, I believe we can turn our hearts to God (as, in the Christian faith, the Bible enjoins us), which makes it easier for us to both deepen our spiritual understanding and to act accordingly. We can and should ready the soil of our hearts for the divine seed to grow and flourish.

To do so is not a contradiction with faith. Rather, it's a responsibility of our side of the relationship. As a Christian, my offhandedly saying that I want "to learn from Jesus" isn't enough. I can't just say that, go back to my normal ways, and expect anything to change in my spiritual life. Spiritual life — *discipleship* — takes work. In other words, simply desiring to be a better disciple and "powering through" isn't sufficient; we can't give up there. We should explore better ways to live out our spiritual beliefs and put our faith into practice. Sometimes that means being wise about our own limitations. We can recognize, for example, that we might need to be reminded to pray, to get encouragement from fellow believers, or to get help removing other obstacles that stand in the way of living full spiritual lives.

This book offers an approach to thoughtfully working around our limitations, drawing from lessons in behavioral science. The approach, spiritual design, entails intentionally changing our environments to support our spiritual practice. In the end, there's no guarantee, and we can't force a relationship with God. But we can nevertheless work to overcome our daily obstacles, and to prepare the way for a deeper, more meaningful relationship.

A Similar Understanding

> *There's a beauty and a humility in imperfection.*
> — *Guillermo del Toro*

In addition to learning from tools behavioral scientists have developed, there's another important overlap between behavioral science and spiritual practice: that is, how both communities often think about human decision-making, and our ability to choose wisely.

I've found that the core understanding of the mind that many behavioral scientists (including me) have and the understanding found in many major religions is remarkably similar. Behavioral economics arose as a critique of economic theories that described human beings as perfectly rational people who know what they want and pursue those ends effectively. Behavioralists struggled to match those theories with the empirical realities of people who were inconsistent in their preferences, failed to choose the "best" path, and overall acted in ways that seemed irrational.

Researchers developed a new understanding: of people who are limited in their willpower and in their ability to understand the choices before them. People who are internally conflicted between short-term and long-term goals. People who are creatures of habit, and subject to the cues and pressures of peers around them, for good or for ill. It's a picture of an imperfect humanity, but also one of hope: by understanding the causes of our behavior and by making changes, people can do better by themselves and others.

Which, of course, is exactly where many major religions start, and they didn't need decades of academic research to understand this. In my own Christian tradition, humanity is flawed and imperfect, but capable of better. We are at conflict within: we suffer under a false self of egoistic desires, yet also are graced with the Inward Light, the work of the Holy Spirit within. We want to be good and godly, but repeatedly stray. The first five books of the Bible — the Jewish Torah — tell the story of a stiff-necked people who repeatedly fail to follow through on their commitments and desires. Throughout the Gospels, Jesus speaks of our "brokenness." In the Epistles, Paul speaks of the power of the Christian community to strengthen us or to lead us astray.

Spiritual design, like behavioral science, is based on an understanding that we're fallible beings, and we all need a little help sometimes to live out our faith. Our minds wander. We forget things. We procrastinate in our prayers just as we procrastinate in doing our taxes. We falter on our path, and we fail to pay enough attention to our spiritual growth. But we can design our environment, to help us work around our human limitations and do just a little bit better.

3
TAKING ACTION IN OUR SPIRITUAL LIVES

We all stumble in many ways. — James 3:2

It's sunny, you're walking down a familiar city street, and you see face after face passing next to you — each person hurrying to work, to school, to wherever it might be. You notice a woman sitting in a doorway, rocking back and forth. Her leg is bandaged up, and today she's not saying anything except for offering a gentle smile to those who pass her by. She's homeless, and she's in that doorway almost every morning.

Often, you notice her, but you don't really see her. This morning, at least, you see. You reach for your wallet, and you start to open it up. But you remember that you don't have any cash. You're not sure where it went — cash never seems to stay in your wallet. And so you have nothing to give her. There's an ATM a block away, but you have to get to work. Maybe you'll remember on the way back. Maybe someone else will help her out. Either way, you're already ten steps past her.

You've probably all been in there, just like me. Personally, I see her almost every day I go to work in Chicago. I used to see her by the subway station when I lived in Maryland. And I've seen her almost every place I've lived before.

It's an expression of simple human compassion — and for many of us, also a religious duty — to try to help those in need, whether it's through offering support on the street, through charities, or other channels. And yet we fail. Time and time again.

But what if we had money in our wallet, in that moment when her humanity and suffering touched us? What if there were fewer people around who could possibly help? What if we weren't in such a hurry?

When I look at my own actions — my ability to show compassion in moments like these — I have to admit that my will to help is…fragile. If I have money (especially, sadly, if I have small bills instead of large ones), I'm more likely to give. If there are fewer people around, I'm more likely to help. And if I'm in less of a hurry, and in a good mood overall, I stop and consider a bit more. But if the stars don't align, I'm probably going to walk by and not even think about it.

And it appears that that's pretty common. There's a fair body of literature about charitable giving, or when people give to those in need. Some of it is about online or mail donations, but some focuses specifically about real life, in-the-moment assistance — monetary and otherwise.[47] We looked earlier at the often-cited study of seminary students: when those students were in a hurry, they stopped to help less often (60% to 10%). In another study, researchers found that 55% more people donated to the Salvation Army when bell ringers verbally asked for donations instead of remaining silent. However, over 25% of potential donors explicitly avoided the vocal bell ringers by changing

---

[47] See, for example, Karlan and List (2007) on the effect of match rates and donations, and Castillo et al. (2014) charities asking donors to ask their friends to donate online. See Bekkers and Wiepking (2011) for a summary of the literature on the factors that matter in eliciting charitable donations.

the door through which they entered the store in the study, probably to avoid the guilt of saying no.[48]

These studies often show the negative sides of our behavior — the situational factors that *hinder* us from helping others. But we can also turn them around — and look at what *helps* us. And we can look more broadly than just at charitable works. More generally, *why do we take action in one moment, and not another?* It turns out there are some very distinct and common patterns which can help us take action more intentionally.

A Framework for Thinking About Intentional Action: EAST

A few simple principles seem to determine which actions we take, and which actions we don't. There are always exceptions, but an easy framework to remember is this: things that are easy, attention-grabbing,[49] social, and timely are more likely to be done. We can remember this framework with the acronym "EAST."

- *Easy*: easy to do, without friction and without significant mental effort or thought.
- *Attention-grabbing*: draws our attention (in an appealing way).
- *Social*: involves others who are supportive or take similar action.

---

[48] Andreoni et al. (2017)

[49] The Behavioural Insights Team (2014). The original framework from U.K.'s Behavioural Insights Team had "A" stand for "Attractive" — in the dual sense of attracting attention and being incentivized. The term 'attractive' has a very different primary connation in the U.S., so I've expressed 'attracting attention' with 'attention-grabbing' instead. For spiritual action, direct financial incentives aren't usually relevant, and I've addressed indirect incentives (like social incentives, frictions as costs, etc.) under the other areas.

- *Timely*: the prompt for action occurs when we have the bandwidth to act, and the benefits are clear in the moment.

Why do these factors matter so much? It's all based on how the mind makes decisions — that behavioral understanding of the mind we talked about earlier. *Ease* matters because so much of our behavior is automated or guided by the reactive, intuitive part of our mind (where small barriers can readily derail us), and because we economize on mental energy. Things that require work and thought, well, often have to wait until later. *Attention* matters because our minds are simply limited: we can't pay attention to everything. *Social cues* matter because one of the main shortcuts that our minds use when we're not sure what to do is to look at what others are doing. And *timing* matters because, again, our mental bandwidth is limited, and a simple rule of thumb is to prioritize the immediate and urgent over the long term, even if it is more important (this is something we call *temporal myopia*).

Thinking back to the homeless person on the street, the *easier* it is to give money (like when we have money in our wallets versus when we don't), the more likely we are to do so, all else being equal. The more likely we are to notice the person, and give him or her our *attention*, the more likely we are to help. So, for example, we're less likely to help homeless people we don't actually see. When others help, we are more likely to as well; when others don't, we also tend to follow their *social* example. When we're less hurried, or when there is an immediate *timely* need (the person is in obvious distress), we're more likely to help.

If someone wanted to avoid the temptation to turn away from those in need, what would they do? In each of these four areas, they'd look for the obstacle, and seek to remove it. They'd to plan to always have money in their wallets, to be able to give it out when needed and avoid the excuse of not having anything to give. Perhaps they'd make a habit of walking down the street where homeless people hang out, asking for money. They'd choose to put ourselves in a situation where they're more likely to do the right thing, rather than to avoid it. Personally, I think I can do better at this — and I think we all can.

EAST and Our Spiritual Lives

The EAST framework was developed in the context of general decision-making and behavior — i.e., regardless of whether or not the behavior has a religious or spiritual dimension. But hopefully, you can see applications to spiritual life. If you find that there's a spiritual practice you want to integrate into your life, start by asking yourself these questions:

- Is it easy?
- Does it draw my attention?
- Is it social?
- Is it timely?

If it isn't all four of these things, redesign the environment so that it is. That is the heart of spiritual design.

To better understand how to do this in practice, let's dive into the first part of the EAST framework in more detail.

Make It Easier

Making an action easy sounds, well, too *obvious* to be true. But we know intuitively that the harder something is to do, generally, the less likely we are to do it. Like it or not, it's simply more likely that we'll give money to homeless person if we have money in our pockets than if we don't have money on hand. Yet, giving to the poor is one of our Christian duties (as it is in many other faiths as well). So let's not ignore the importance of making things easy.

In the research literature, there are many ways to make an action easier. Here are a few obvious and not-so-obvious approaches:

- Set up defaults.
- Remove "hassle factors."
- Start small and celebrate "small wins."

- Plan out future contingencies with "implementation intentions."
- Reframe the action as a natural extension of our identity and past.

Set up defaults

One of the most central lessons in the behavioral literature is about *friction*: small hurdles can stop us from taking actions that we otherwise would take. An early (and startling) example of this comes from the realm of organ donations. Organ donations are an ethically important subject. We literally have the potential to save someone's life. However, there are huge variations in participation in organ donation programs across countries, with many countries either having 98-99% of the population agreeing to donate their organs upon death, while in others, only 0-10% plan to do so. Even neighboring countries with similar histories and cultures — like Germany and Austria — show these variations: Germany has a 12% rate; Austria has a 99% rate.

The reason for these differences isn't because of a deep-seated ethical or religious understanding of organ donation. It's likely because Austria defaults people *into* their organ donation program, and lets them readily exit if they choose. Germany defaults people *out*, and lets them readily enter if they choose. It seems that what matters is the simple act of checking (or unchecking) a box on a form. That's the incredible power of a small friction (merely checking or unchecking a box) and the default presented to people.[50]

Another less vivid but still important example comes from the retirement savings world. Researchers found that defaulting people *into* their workplace retirement savings plan (and letting them leave if they want to) nearly doubles the percent of people who contribute, versus defaulting them *out* and asking them to join if they want to. In various surveys people claim they want to save, but the minor friction of checking the box stops them.

---

[50] Johnson and Goldstein (2003).

How might we set up defaults *for* spiritual action? Well, drawing directly on the example of retirement savings, if one's goal is to tithe at church or to give money to the needy, we don't have to rely on having money in our pockets. We can set up a smart default with automatic contributions from our checking accounts or, even better, a paycheck deduction that occurs even before money hits our checking accounts.

How else might defaults support one's spiritual life? Consider the radio station in your car: is it set by default to a spiritually inspiring station? Have you signed up for emails that automatically provide guidance in spiritual growth and scripture readings? Do you choose a daily commute that takes you past a church or temple?

Remove "hassle factors"

There are other ways beyond defaults to remove friction from one's spiritual life. Researchers call smaller frictions "hassle factors": the small annoyances that slow us down and give us reason to procrastinate or avoid taking action. For example, for many families who try to fill out the FAFSA student aid form,[51] the process is full of hassles which stop them from applying despite the significant amount of money at stake. One research study gave a randomly selected set of students assistance with the FAFSA, which increased the likelihood that they would complete it by 15.7%.[52] Seemingly minor hassles indeed can lead us astray.

In terms of spiritual actions, consider the act of joining a community of believers for church services or meetings. What small annoyances might get in your way of this? A lack of parking? Getting the kids out the door? Feeling like you have to dress up and not having the clothes you need? Rather than trying to power through and saying to yourself, *If it were important enough to me, I'd do it anyway,* you can work out a plan to remove those hassle factors. For example, you could get a ride with a friend to avoid the frustrating search for parking. Would it be possible

---

[51] Free Application for Federal Student Aid: the federal form in the U.S. that parents and students use to apply for college loans and aid.

[52] Bettinger et al. (2012)

to get everything you and the kids need ready the night before? Or make an informal agreement with your friends at church so that *none* of you dress up?

Behavioral researchers have studied and documented the importance of these hassle factors. But you don't need to be a behavioral scientist to fight them. Figure out the things that are standing in your way, no matter how trivial, and come up with creative ways to remove them.

Celebrate small wins

Have you ever felt a sense of satisfaction — or even excitement — when you cross something small off your to-do list? Researchers have studied what are known as "small wins" and how they can help us gain momentum over time. It works like this: if you want to do something big, start with an achievable piece from which you can see a clear accomplishment. The initial research on this was conducted with credit card payments. People that pay off small balances first, even if those balances are not on the highest interest cards, are often better off. We can use the idea more broadly, however.

In trying to go to church every week, don't think about "every week." Instead, go once, and celebrate that accomplishment. Or successfully get everyone out of the house on a Sunday morning, even if that time isn't then spent going to church. Going to church regularly comes after you've made, and celebrated, the smaller progress towards your goal. To increase your giving, start by donating whatever pocket change you have on a single day to the Salvation Army or a similar group — and build up to more meaningful contributions over time.

Plan out future contingencies with "implementation intentions"

Researchers have found that after we've committed to take a particular action, obstacles that arise along the way can make us stop and think about what to do next. In those moments of pause, we can get distracted and stop altogether. One of the solutions to this is what's known as **implementation intentions**: planning out beforehand what might happen and what to do about it. Thus, when a problem arises in the future you don't have to stop and think; you already know what to do.

Here's how it can work. Let's say you've had a falling out with an old friend and decide to meet up to grant forgiveness. Don't just leave the decision at that. Instead, think through potential obstacles:

What will I do if, when I reach out to them, they don't want to meet in person?

What will I do if we can't find a time that works for both of us and we start to argue over that?

What will I do when we actually meet in person?

What are the actual words that I will say?

The goal is not to obsess over potential problems, but rather to work through the next steps in case of these contingencies. That clears the path ahead to do as you intend.

Reframe the action as a natural extension of our identity and past

Let's say you've felt a calling: to organize a discussion on a pressing issue for your spiritual community. But it's something you've never done before, and you're afraid or uncomfortable. Researchers have studied how powerful it can be to reframe an action as an extension or progression of things we've done in the past. By connecting it with our prior experiences and who we see ourselves as individuals, we can feel more confident. Tim Wilson's work in particular, including his 2011 book *Redirect*, focuses on the reframing of our past experience to make future success more likely. James Clear, in his book *Atomic Habits*, describes a related approach, in the context of identity-based habits:

> Your current behaviors are simply a reflection of your current identity. What you do now is a mirror image of the type of person you believe that you are (either consciously or subconsciously).

> To change your behavior for good, you need to start believing new things about yourself. You need to build identity-based habits.[53]

For example, when you're faced with an unfamiliar and anxiety-inducing spiritual leading, like directing a community discussion, think about other discussions you've led outside of your spiritual community, or other things you've organized in your daily life — from a trip with friends to more complex organizational hurdles at work. Rather than focusing on the *unfamiliarity* of the discussion in your spiritual community, focus on that which is *familiar*. Look at how it's gone well for you in the past, and how that's an expression both of who you are and the skills you have as a person.

A Few Lessons

How might we apply "Make It Easier" to our spiritual lives? We've talked about some specific tips above, like setting defaults and thinking though implementation intentions. But how do you figure out where these are needed?

You can use a short process to find the problem area — that is, what's stopping you from acting — and then try out creative solutions. Here's the process:

1) Look at what you want to accomplish and *find clarity within* on whether that is the right goal. Understand where you are led through discernment.
2) *Map out what you need to do at each step of the way.* Instead of, "I need to get a better car to deliver meals for charity," think, "I need to save X dollars to afford a new vehicle." Next, "I need to go to the car dealer on Y day and research my options." Then, "I need to buy a car with extra storage in the back." The more specific you make the actions, the better.

---

[53] Clear (2018).

3) *Look for small frictions along the path.* Where does the process annoy you? Intimidate you? Where do you tend to get distracted?
4) *Try changing your environment to remove these frictions.* Don't rely on willpower alone!

In fact, this approach can be used for any of the behavioral obstacles to action we discuss in this book, from making it easier to devoting our time and attention to spiritual practice.

4
DEVOTING OUR TIME AND ATTENTION

...how is it that you are turning back to those weak and miserable forces? Do you wish to be enslaved by them all over again? — Galatians 4:9

Have you ever wondered why it's so hard to put down your phone? Or to stop playing a game? In part, it's because people like me have designed them that way. We've helped design products and marketing campaigns that surround you in your daily lives, capture your attention, and keep you hooked over time.[54] Consumer psychology and behavioral science are often baked into their design to make them more

---

[54] For example, you can read a summary of product psychology — designing products to be compelling and even addictive at www.nirandfar.com/best-of-top-picks and www.productpsychology.com. For good or for ill, members of my team and I helped develop some of the training material on the latter site.

appealing, more shareable and more all-consuming of your time and focus.

Sometimes these techniques make the products more valuable for their users. Sometimes they simply help the company selling them, like Netflix's or Amazon Prime Video's go-on-to-play-the-next-episode technology that sends you automatically into the next episode of a series you're watching in a few seconds unless you choose otherwise. Either way, these products and marketing campaigns make it harder for you to make time for spiritual practice — or even to find time for meaningful communion with others. We're all guilty, at one point or another, of spending too much time on an electronic device to the detriment of those who matter most to us: God and our loved ones.

If devoting more time to spiritual practice — or spending meaningful time with those you love — is something you'd like to do, then making it easier to do (which we covered in the last chapter) is a good start. However, it's not enough. You should *attack the problem of attention head-on* by making the things you want to do more likely to grab your attention and removing distractions that otherwise would consume you.

Grabbing Your Own Attention

As with "making it easy," there are some obvious ways that researchers have studied to get attention to an action, and there are nonobvious ways which can also help you when you struggle. Here are a few of those obvious and nonobvious techniques.

Set up clear and powerful cues

Growing up in my family, I didn't think much about religion. It wasn't something my parents talked about, nor was it something we saw or interacted with. And so, given the wandering mind of a young child, it's understandable that my mind wouldn't turn to something I wasn't aware of in the first place. Later, in college and during much of my professional career, I was simply busy with other things. It wasn't that I *rejected* church and religion, it was that I simply wasn't *thinking about it* in the first place. No reason I would.

In the behavioral literature, our attention is guided by two major forces:

1) Cues from the outside world (external factors), and
2) Internal states like hunger or fatigue.

There are literally an infinite number of things we could be thinking about at a given moment, and our minds don't magically pick "the most important thing." Instead, we're built to *react to our environment*. We see a tiger running towards us, and we suddenly think a whole lot about tigers and how to escape from them.

In daily life, our environment grabs our attention and focuses it on particular things, whether that environment is a conversation with friends, an email we receive, a video we watch, an ad we see, or something else. If there isn't something that *cues* us to think about church and faith, we're less likely to do so.

In my case, for much of my life there simply weren't many external factors directing my attention to God and church. However, when I moved back to the area where I grew up and drove past the old Meeting house I attended, it reminded me about the peaceful calm I had when I was there. In other words, it brought the issue to my attention and triggered me to act.

Like many religious people, I believe we that have a built-in yearning for meaning and for connection with the Divine which can draw our attention. Some people liken our internal yearning to a need or a hunger. However, this hunger doesn't seem to guide us to a particular action (like prayer or meditation) unless we've already learned how to meet that need. And for myself, I hadn't experienced that connection nor had I learned tools to help seek it out. I felt there was something missing in my life but I'd never experienced anything that could fix it — and so that internal hunger couldn't focus my attention on a specific action.

Use reminders

One of the most powerful reminders I have to turn my thoughts to God is the cross I bought in a church gift shop many years ago. I leave it on my nightstand each night. I see it in the morning when I dress — because it's right in front of me — and so I put it on. When I wear it, it helps me bring my attention to my faith and spiritual goals. Every time it moves, or when I see it around my neck, I'm reminded.

Prominent behavioral scientist Dean Karlan of Northwestern University and others studied how simple reminders can overcome our *in*attention and return our thoughts to the things we intended to do. For example, they gathered together a set of people who wanted to save for the future, and randomly assigned them into groups: one group received text messages reminding them about their desire to save, and the other didn't. The group that received the text messages saved 14% more than those who didn't.[55] It's not that the latter didn't know they wanted to save (they'd indicated they did in the first place!); it was that they simply *forgot*.

Look for times when you'll be most receptive

There are times when we're overwhelmed with information and with demands for attention, and there are times when we aren't. During the overwhelming times, we can try to overcome all of these competing demands for our spiritual focus — or we can be wise and avoid them by waiting.

Let's say we wanted to read the Bible more often. We intuitively know that trying do that while we're driving is probably a bad idea — for safety reasons of course, and because our minds are otherwise occupied. Yet, we don't always apply that lesson in other, less extreme situations.

For example, trying to read at night before going to bed can be a very difficult time for some people because they're exhausted and unable to

---

[55] Karlan et al. (2010)

concentrate. For them, the best time to read might be while riding the subway home, when there's a window of time between work and home. During this time, they're in transition and relatively free from the demands of each of these facets of their lives. Personally, it's on the ride home in the afternoon that I've found I am most receptive to reading spiritual works.

Go to where your attention already is

A more nuanced and clever way to remind yourself of something you want to do, is this: figure out where your attention will normally be at a particular time of day and put a reminder there. Then determine how you can make it possible to take the desired action right then and there.

Marketers have used this trick for a very long time. If you've ever received pens or a water bottle with a company's name on it, or preprinted home address labels in the mail, that's what they are doing. Those items aren't simply reminders of a particular company or charity, they are reminders designed to go where your attention will already be: on mailing, on writing, on drinking water.

What does this look like in a spiritual context? Some options we intuitively know: signing up for a daily Bible message via email; placing a cross prominently in the kitchen; placing the religious books on the bedside table; putting a calendar with a daily spiritual message or thought on our work desk. We might also install a spiritual app on our phones on the first app screen (where we'll often look each day), or have it provide a daily message or prompt for prayer as our background image or locked-phone screen. Each approach builds on where our attention is, rather than trying to redirect it to a new and unusual place.

Build on existing habits and routines

Social Psychologist BJ Fogg, head of Stanford's Persuasive Technology lab and pioneer in the field of intentional behavior change, has a nice framework for thinking about starting new habits. He argues that in order to start something new, figure out what routines you already have. Then identify the end of an existing routine, and tack on the thing that you want to do next directly afterwards. You seek to build a strong association in your mind between whatever it is you're currently doing, and the new thing. Look for where your attention (and behavior) already is, and build upon that.

For example: whenever I finish brushing my teeth, that's when I need to use mouthwash. Or in a spiritual context, when I put the kids to sleep, that's when my spouse and I sit down and talk about the ways that we are grateful for God's presence in our lives that day. The idea is to go where our attention already is. Form a link between where your attention currently is (the current routine) and where your attention should go next (the thing you want to do).

Each of these techniques — visual or external cues, reminders, getting in the line of sight, and building on existing habits — can help us by pulling our attention *to* our spiritual practice. That alone is often insufficient though. We also need to address the myriad forces pulling us *away*.

Removing Distractions

> *It's easier to be a monk in a monastery.*
> —*Traditional Buddhist Saying*[56]

If we were really sincere in our religious pursuits, wouldn't that be enough? Wouldn't we be okay, somewhere between the power of our efforts and God's grace? In the abstract, sure. When we look at the daily details of our lives this logic fails. Personally, I yearn to live a more contemplative life. My deepest peace, and foundation of joy, comes from staying in silence, seeing God's Creation around me, and feeling my connection to it.

---

[56] Many thanks to Ryan Murphy for sharing this teaching.

But I have a job. And kids. And friends. I have things I need to "do." Each can serve as distractions from spiritual pursuits, and limit our ability to find God's peace within.

Some of these distractions can, if approached well, help us "love each other as we love ourselves and God." When I feel the peace within *and* play with my kids at the same time — now that's a special feeling. Or, as Merton writes about the distractions of work:[57]

> To do work carefully and well, with love and respect for the nature of my task and with due attention to its purpose, is to unite myself to God's will in my work.

Many other distractions *don't* build our relationships with one another or with God, however. First among them are the thousands of ads we see every day.[58] Ads meant to make us feel inadequate. To make us aware and ashamed of a hole in our lives that isn't real. Or ads that encourage us to define ourselves through what they sell. To encourage us to take pride in our purchases, instead of what we accomplish for others — or instead of avoiding pride altogether.

The makers of these ads are not bad people. Rather they are doing their jobs, jobs that may be as much a part of God's plan as our own work is. I am not wise enough to judge the person, but I can certainly judge the outcome on my and other's lives though.

We are surrounded by appeals to *buy something better*. But we are surrounded by very few appeals to *be better people*.

To the best of my knowledge, there hasn't been behavioral research on how to best remove distractions — unlike the body of work just

---

[57] Merton (1961/2007), page 19

[58] See Story (2007)

Spiritual Design

described on how to grab attention.[59] But we can make a few inferences from the research, and from daily experience, that may help.

- *Remove distractions — don't just ignore them.* If you try to ignore the things that normally distract you, that takes effort and time, so it's often wiser to change your environment so that you're less likely to be distracted.

  You can remove common distractions from the places you normally look. For example: if you're distracted by your phone at home and don't spend the time you'd like to with you family, drop your phone in a box when you enter the house. If you're distracted by an app that sucks up your time, either remove the app, or make it harder to get to (move it to a different, later screen page). In other words, remove the cues: the things that remind you of distracting activities.

- *Crowd out distractions.* Look for the times in which you find yourself pulled into activities that are momentarily interesting, but ultimately unsatisfying. For me, it's late-night Internet TV — after watching I often feel numb and am left with less than I started. Instead of simply telling yourself NOT to do that, enumerate and facilitate alternatives. What would you enjoy in that moment that also feeds you within? Have those options ready when you feel the dull boredom ("Well, what else am I going to do now?") that leads to empty distractions.

  For example, place things on or near the couch (books, games, etc.) that can be an alternative to mindless TV watching. Apps on the phone, ironically, can also be good alternatives when well chosen: Duolingo is one of my favorites. Similarly hiking boots placed near the door, or a comfy chair on the porch, can help ensure you always have something more rewarding to do than watch TV.

---

[59] Though interest is growing in this area. See Eyal (2019) for a set of guidelines on how to control attention. Wendy Wood, a leading habit researcher, also has tips on breaking habits like looking at your phone – see Wood (2019).

- *Beware of "the normal."* The things that grab our attention do more that distract us: shape our expectations of normal behavior. They create descriptive norms:[60] our understanding of what other people are normally doing which shapes what we do ourselves. Descriptive norms are great if they demonstrate a loving community of believers (something we talk about in the next Chapter); not so much when they show us it's normal to constantly feel inadequate unless we have the fanciest new stuff.

We can't avoid descriptive norms altogether, but we can set more meaningful and helpful ones — which is something we'll dive into in the next chapter.

A Few Lessons

Marketers, product designers, and behavioral scientists are all trying to capture your attention. That means you're not facing a blank slate on which you can fashion an ideal environment to promote your spiritual practice. You're starting in an environment that already is designed to build habits of distraction and, along the way, lighten your wallet. You're starting with the scales tipped against contemplation and non-material pursuits. It's hard to remove the distraction from our lives — it's hard to be contemplative — because no one makes money off of our contemplation.

Spiritual design is a small attempt to balance the scales: To recover the power to direct your own attention to what you find important in your life. And you can do that when you:

- *Make explicit cues for practices and mental states you want to engage in,* such as the Bible-by-the-bedside.

---

[60] See discussion in Chapter 1 and Gerber and Rogers (2009).

- *Use simple reminders*, such as calendar items (on your phone or on the fridge) to give time for God, especially at a regular time of day.
- *Look for where your attention already is* — on your laptop's desktop screen or browser homepage, or on your bathroom mirror after you finish brushing your teeth — and create ways to think about and act on your spiritual practice in that moment.
- *Remove distractions* — on your phone, on your computer, in your physical environment — so that you have more time and attention to spare for the things you care deeply about.

5
THE COMMUNITY AROUND US

The Christian needs another Christian who speaks God's Word to him. He needs him again and again when he becomes uncertain and discouraged, for by himself he cannot help himself without belying the truth. He needs his brother man as a bearer and proclaimer of the divine word of salvation... And that also clarifies the goal of all Christian community: they meet one another as bringers of the message of salvation."

—Dietrich Bonhoeffer, Life Together

The Gift of Community

Humans are intensely social beings, and most of us need to interact with others in order to be happy. That's true even for those of us who are introverts (like me): We just want to interact with others on our own terms. In many ways, we are wired to pay attention to and focus on social interactions.

One of the most fundamental lessons from behavioral science (and across psychology) is that we look to others for cues about appropriate behavior. In other words, we have an inclination to "follow the herd." As I've mentioned prior, our sense of what is normal, and our own subsequent behavior, is shaped by those around us. This is no secret, and it's not an insight that is unique to behavioral science.

What's more interesting, though, is how behavioral science teaches us what "community" really means, and how we can apply its lessons to use the power of community to help us on our path.

The community that affects our behavior the most isn't necessarily a fixed or obvious thing. Instead, it is often selective and constructed. One can think of the group that shapes us and our behavior as a **reference group**: the specific people we compare ourselves to and learn appropriate behavior from. Our reference group may align nicely with the people who live near us, but it also may not. Niche communities on the Internet have made this clear: someone with a particular political or personal interest can find acceptance and encouragement (or radicalization) among like-minded people online, even when their physical community is hostile to that interest.

Often such niche Internet communities are seen in a negative light. However, the same dynamics of social support occur in beneficial ways in other online and online communities all the time: in members of sports teams or study groups pushing each other to excel, for example, and so on. When you understand that reference groups are constructed and not given, you can help support your own spiritual growth as well.

Let's look at some research lessons that could be translated into a spiritual context, to help support faith and practice.

Set a New Reference Group

If we aren't sure what to do in a particular situation, we often look at what other people are doing and try to do the same. Following a descriptive norm like this is one of the most basic shortcuts that we use to guide our behavior in everyday life (and something which we discussed briefly in Chapter 1). This shortcut can help us get to where we desire to be... or, if we're not careful, it can hurt us.

To understand how this works, think about the people you see in your daily life. Who do you implicitly or explicitly look to for cues on how to live? Are they living the deep and meaningful life that you seek? Are they expressing loving kindness to the people around them, and to God's creation? Or are they caught in a trap of forever grasping for more, and never being satisfied, never being whole? I think for most people, it's the latter. When we're confronted with these questions, we can usually think of one or a few people who do live in this way — but sadly those aren't the people we normally measure ourselves against. Instead, we model our behavior after the constant search for status and stuff that we see in others around us. In other words, our descriptive norms hurt us. But it needn't be that way.

Let's say you want to pray before each meal. One way to support that action is to intentionally seek out others who show their faith in that manner. You then find a setting in which people already pray at mealtime. The shortcut then helps: "everyone here prays at meals; it's okay for me to do so too." More generally, descriptive norms are one reason why going to church can be so powerful: we are surrounded by people who show us that spirituality is normal and provide examples we can draw from on how to live a spiritual life.

Another way to use these descriptive norms is more abstract: to look for statistics about people in similar circumstances. In the U.S., despite all of the doomsday rhetoric, the vast majority of people believe in God[61] and live according to their faith in one way or another. Reminding ourselves of these facts can help us remember, once again, that spirituality is "normal."

On a personal note, the challenge I have (and likely others do too) is that where I live and work, almost nobody seems to pray at meals.[62] If I were to look around at mealtime to see what other people were doing,

---

[61] Pew (2018)

[62] Which may be because they pray, but make a point *not* to show it — see discussion on Matthew 6 in the final section of this chapter.

the shortcut of descriptive norms might backfire: "I don't see anyone praying before their meals here, so maybe neither should I." However, since I already know that others aren't praying, I intentionally look down and focus on my food before I start praying. That helps me avoid the negative power of conformity.

Join Others or Create the Community

> *How good and pleasant it is when God's people live together in unity!* — Psalm 133:1

In addition to descriptive norms that show us what's normal from a distance, a close connection with a supportive community can be very powerful. First, social expectations are powerful on their own because we tend to want to meet those expectations, all else being equal. According to brain scans, going "against the crowd" can activate fear responses in the brain.[63] Clearly, when the group we're with is behaving badly, the fear of going against the crowd can get us into trouble: that's negative peer pressure. Our natural tendency to take cues from those around us isn't inherently bad however. We can use it to our advantage by finding a supportive community: i.e., we can seek out *positive* peer pressure and use it to support what we already want to do.

Personally, I've refrained from drinking alcohol for most of my life. At parties where many people were drinking, I intentionally sought out those who weren't. We created a small group of people who self-identified as nondrinkers. I found that simply knowing that there were other people in my community of nondrinkers helped me resist the broader peer pressure to drink.

Community means more than peer pressure, of course: community members can directly support each other with encouragement and assistance when they struggle to take faithful action. In my case, I

---

[63] Berns et al. (2005)

remember a time when I was at a party and a fellow nondrinker saw that someone was pointedly asking me why I wasn't partaking. The person stepped up, and intentionally showed that she wasn't either — to support me in the moment and show solidarity.[64]

There's another aspect of community and social interaction that can help us follow a spiritual path: *often taking faithful action alongside other people is simply more fun.* We may simply enjoy talking with others, or sharing experiences with them. Similarly, when there's something we're trying to do but are struggling, we can try inviting others to make it more fun for us. There's nothing wrong with that — it's how many good church groups, social movements, and charitable organizations get started: with friends. So that's another obvious but sometimes forgotten technique: *invite friends.*

Raise the Stakes and Create Social Commitments

If you tell yourself that you'll go to Bible study next week, that's one thing. If you call up your friends and tell them to expect you there next week, that's another. If you don't go through with what you've promised, imagine your friends' disappointment with you. You've effectively raised the stakes for yourself to follow through and do it. One of the ways in which we can help ourselves act according to our faith is to make *social commitments* to do so — like telling your friends you'll be at the next Bible study.

By making a social commitment, we set expectations in the minds of others about our behavior. We naturally have a tendency to want to be consistent in our actions, and we especially want to do so when others are watching, and we know that our reputation with them is on the line.

There are many ways in which we can create social commitments. An obvious and simple one is to *tell people we care about what we plan to do.* For

---

[64] Thank you again, Eirin.

example, tell your parents that you plan to start tithing when you get your next raise.

We can also go a step further: *ask our friends and others to hold us accountable* if we don't follow through. For example, I have a friend who's struggled with a behavior that isn't in line with his faith. He knew that it was going to be difficult to change, and so he asked God for help. He also asked his friends. He put a plan in place to do better, and then he asked his friends to check in on him, and to hold him accountable if he didn't stay the course.

Another way in which social commitments affect us is implicit: *through the type of people we say we are, how we demonstrate ourselves to be, and how others expect us to act.* We often have a sense about how others view us; if that view is positive, we may be loath to undermine that positive view. We not only want to be held in high esteem,[65] we don't want to fall short of the type of behavior that others expect of us.

Personally, I want to be a good father to my children, and I want to teach them to be decent, loving people. I know that starts with my demonstrating that to them in my own behavior; my words aren't enough. They expect me to show the decency that I expect of them. Usually, I think I do okay. But every time I get upset at them, every time that I don't listen properly to them, I know that I'm not only doing wrong by them, but I'm falling short of how they view me. I'm changing how they'll see me in the future — which is a pain that stings like nothing else in this world. We are role models for our children and others, whether we like it or not.

Embrace Your Spiritual Identity

One "superpower" of being part of a community is the identity we share with others. That identity guides us when we're unsure of what

---

[65] A broad psychological mechanism with many other ramifications that I won't touch upon here.

to do. When we're faced with a dilemma, we ask not only, *What should I do?*, but, *What should I do as a Christian (or Buddhist, etc.)?*

Christianity, and indeed all religions and spiritual groups, provide guidelines for how we should live our lives. These guidelines can help us find the moral path when we're tempted to do otherwise or simply unsure of what to do next. However, that only helps to the extent that we personally *identify* as that type of person. And the stronger the self-identity — and the more accessible that identity is to us in moments of uncertainty — the more likely it is to help when we need it. One way to reinforce that identity is to think about what you "call yourself" and, if you feel comfortable, refer to yourself in that manner publicly as well.

Another powerful way is to *formalize your membership at your church or other spiritual community*. Even if you don't attend church more often, it helps make real your identity and that you are bound together in a common relationship to God with others in your community, through good times and bad. Mentally changing from "I'm here to check things out and see if I like it" to "I'm a member of this community" matters. It reinforces your social identity as a believer and a member of the community of believers.

Since high school, I've identified as a Quaker. I've found great comfort in the example set by Quaker friends and Quakers through history — examples that inspire me to follow a sincere and spiritual path myself. However, one of the areas that I've faltered on my path to God was not reinforcing that social identity.

For example, my wife Alexia and I never established membership in our old Quaker Meeting in Maryland. It was a very old meeting, filled with weighty and thoughtful Quakers — people who expressed their faith in eloquent words, but even more eloquently through their silence and actions. She and I knew that becoming members would bring us closer to the Meeting and to our own spiritual path. But since we could interact with the community in a casual manner either way, we didn't take the formal step of becoming members. We started the membership process, but never quite got around to completing it and handing it in. We procrastinated. When we moved to Chicago, that opportunity was lost.

Now that we are part of a wonderful Lutheran Church, we are similarly surrounded by thoughtful, caring people — people who are bound together in Christ and in shared community. I'm glad that we've learned our lesson: my wife and kids have become full members of the church, and I have become a fully committed and active participant in the community.[66]

A Few Lessons

In daily life in the U.S., it is easy to get the impression that most people aren't religious (or otherwise spiritual), for a variety of reasons. First, we have social taboos against talking about religious matters in public. And since other people don't talk about God, we might assume that other people aren't *thinking* about God either. According to anonymous surveys though, that's not the case. In the Pew Foundation's latest research, 90% of American adults believe in a higher power; 80% believe specifically in God; and 56% of American adults believe in God as described in the Bible.

If people believe, then how can we get the impression that they don't? In part, it's because people are *afraid of talking about God*, for fear of social judgment.[67] And because of that lack of public discussion, even more people become fearful of talking about God, etc.

---

[66] Theologically, I'm a Quaker, and it would be dishonest to claim otherwise and request formal membership at our Lutheran Church. As much as I'd want to be an official member, and be more closely bound to the church, I am unable to take my own advice in this area.

[67] Merritt (2017)

Many religious people also don't outwardly demonstrate their faith *because of their faith*. As Jesus teaches Christians:

> *Be careful not to practice your righteousness in front of others to be seen by them. If you do, you will have no reward from your Father in heaven.*
>
> *So when you give to the needy, do not announce it with trumpets, as the hypocrites do in the synagogues and on the streets, to be honored by others. Truly I tell you, they have received their reward in full. But when you give to the needy, do not let your left hand know what your right hand is doing, so that your giving may be in secret...*
>
> *And when you pray, do not be like the hypocrites, for they love to pray standing in the synagogues and on the street corners to be seen by others. Truly I tell you, they have received their reward in full. But when you pray, go into your room, close the door and pray to your Father, who is unseen.*
>
> *– Matthew 6:1-6*

Certainly, this injunction isn't followed uniformly, but with many sincere believers you wouldn't know about their faith unless you purposefully spoke with them about it, or you inferred it from their actions.

At a societal level, you may think it's a good or a bad thing to avoid religious discussion[68] and for people to avoid obvious outward expressions of their faith. But at a personal level, there are unintended consequences: the 90% of Americans who believe in a higher power

---

[68] Personally, I think that overall our taboos hurt us in this area, even if talking about God more openly would entail some uncomfortable situations.

can nevertheless think they are alone, or in a small minority. This can limit our own search for God and spiritual growth.

By understanding the behavioral literature on social cues, we can counteract this challenge in our own life. For example, we can:

- *Carefully select the people to whom we compare ourselves* and look to for guidance when we're not sure what to do. So if you're trying to express your faith in a particular way — whether it be giving to others, praying or simply reading Scripture — look for a community or other reference group where that expression is normal.
- In addition to simply referring to communities where a faithful expression is normal, we can *join or create such a community*.
- We can *"raise the stakes" to breaking our personal commitments by making them social commitments*. We tell our friends, family, or others of the spiritual commitments we've made or are making, to keep us "on course."
- We can reinforce our path not only by getting involved with spiritual communities, but recognizing that we are part of them ourselves, and *embracing the spiritual identity that comes with them*.

6
FINDING THE URGENCY TO TAKE ACTION NOW

Never put off till tomorrow what may be done [the] day after tomorrow just as well.
—Mark Twain, "The Late Benjamin Franklin"

Christianity has a problem of urgency. *Why pursue a more faithful, spirit-led life right now, when God loves you and will welcome you at any point in the future?* Why change now, especially if that means making hard choices, altering one's routines, and facing one's past with honesty? Wouldn't it be nicer just to wait until tomorrow?

In many branches of Christianity, God is always available and welcomes those without faith and inveterate sinners into the fold. Which makes it rather difficult to make the argument — to ourselves or to others — that there's an *urgent* need to change our ways now. From a behavioral perspective, God's grace can be seen as a license to procrastinate.

The same can be said of some branches of Buddhism. It takes a lifetime or more to achieve Nirvana. It won't matter in the scheme of things whether you work at it today, or wait until tomorrow. And likely the same holds true in other traditions with which that I'm less familiar. To see the problem in more depth for Christianity, let's take a brief detour into theology.

The Prodigal Son

Consider the parable of the prodigal son. For those who aren't familiar with it, it can be found in Luke chapter 15, and here is a short summary.

The parable is about a father (God) who has two sons. The younger son is reckless and wasteful (i.e., "prodigal"), leaves town, and squanders the money his father gives him. The older son is faithful and responsible, and works alongside his father at home. Over time, the younger son runs out of money, and returns to his father to beg forgiveness. His father welcomes him, and celebrates his return — while his older brother fumes at the apparent injustice. His father reproaches his oldest as he shows his youngest boundless love and mercy in welcoming back the son who had been lost to him.

While the parable illustrates God's mercy and forgiveness, it also illustrates the problem of urgency: the son who stayed with the Father from the beginning ended up being no "better off" than the one who lived a life of excess, and then returned. The Scriptures are filled with such examples of unmerited grace. The repentant sinner is welcomed by God as much as — and perhaps even more— than the life-long saint.

How does one handle this perverse incentive to ignore God's call for as long as possible, and have some fun in the here and now?

Or, as Augustine so nicely put it:[69]

> But I was an excessively wretched young man... when I actually begged you for chastity by saying, 'Give me chastity and self-restraint, but don't do it just yet'. I was afraid that you'd hear my prayer and quickly cure me of the disease of lust, which I preferred to have satisfied rather than nullified.

Yet to say that there is a theological problem doesn't mean there aren't already theological solutions that others have identified. We can find one solution in other stories of the Bible, such as the parable of the ten bridesmaids[70]: since we don't know when Jesus will return, we should always be ready by living a life of faith here and now. And indeed, some branches of Christianity place tremendous emphasis on the need to be ready — with a righteous fear of the imminent end of the world.

There are many other ways to handle that problem — both from a theological perspective, and, more to our interest here, from a practical perspective in our own lives. The theological challenge is often framed as one of belief or bad behavior: where the person doesn't want, or doesn't feel the need, to change. However, the same issues occur for those who sincerely want to do something differently in their spiritual lives as well, whether to meditate regularly, engage with God's word, or show loving compassion to others. It's the nature of human beings that we are distracted and conflicted: we want and are drawn to multiple things. And those things that aren't urgent often lose out.

When we want to live a more faithful life, but we're conflicted with other desires and interests — especially when those other interests are immediate and seemingly urgent — *what do we do?* To find some answers, let's talk about a very different domain of life: retirement savings.

---

[69] Augustine (2018 translation): Book 8, Chapter 7, 17.
[70] Matthew 25:1-13.

Why We Don't Save

Retirement savings is one of the most heavily researched areas in behavioral science.[71] Saving for retirement has a problem of urgency (to put it mildly). You need a certain amount of money in the future to live a comfortable life in retirement. At each moment of your life, you have the choice: *do I spend the money I have now, or do I save some of it for the future?*

In this choice, you are subject to a variety of biases, including:

1) **Present bias.** We overweigh the importance of the present versus the future, in part because our mental representations of the present are vivid and clear, and those of the future are vague;
2) **Uncertainty bias.** The future is full of uncertainty, and that uncertainty makes it all the easier to postpone action and ignore until later; and,
3) **Delaying pain.** Saving for the future entails, on the surface, an immediate and known loss, in exchange for a long-term and uncertain gain. People generally avoid unpleasant and painful activities, especially if the benefit is uncertain.

The things you could do right now with your money are vivid, real, and immediate, like buying gadgets, paying the rent, and having a good time. The things you can do the in future are vague, uncertain, and far off, like paying medical bills for illnesses you may never contract, or giving money to help your hypothetical future children (you might see some other parallels with faith here too). Retirement savings is often presented as a losing proposition: certain suffering now, for potential future benefits.

---

[71] See for example, Thaler and Benartzi (2004); see Fertig et al. (2018) for a broad, three year-long field study of techniques to increase retirement savings.

And, not surprisingly, in general most Americans do not save enough for retirement — at least not via voluntary, intentional decisions.[72] So, what can help us overcome this behavior? Researchers have explored a variety of techniques we've already covered in other chapters, such as *removing frictions* to make saving easier, *using reminders* to draw attention, and *making peer comparisons* to leverage the power of community. Some of this research specifically addresses our lack of urgency, and the techniques they've discovered to counter it are of interest to our discussion:

- *Make the future more salient* — for example, by having people see age-progressed versions of their face, or visualize their future in retirement — so old age can become more vivid and meaningful to them, so they will be willing to better prepare for it.[73]
- *Use metaphors* and situations people can readily relate to and understand from their daily lives. For example, don't talk with people about how much money they will have in retirement. That is often a large number, and even if insufficient to live on, can create what's known as the Illusion of Wealth.[74] Instead, show how much they'll have to live on each month.
- *Create a specific moment* in which people are forced to make a decision (in the studies, the decision was how much to save for retirement), and then *automate the process* until the next time the person chooses to engage.[75]
- In a moment of strength, *commit to a desirable change and make it difficult to reverse course in future times of temptation.*[76] For example,

---

[72] There are a variety of other (nonconscious or non-voluntary) tricks that governments and companies have used to promote retirement savings; they aren't the focus here.

[73] Age progression: Hershfield et al. (2011); Visualization: Hershfield et al. (2018)

[74] Goldstein et al. (2015)

[75] Keller et al. (2011)

[76] Ashraf et al. (2006)

by putting money in an account that you can't withdraw from until a specified future date.

Some of these techniques are specific to retirement savings (like age-progression), yet there are others we can learn from and apply in our own spiritual lives — especially those mentioned in the last two bullets. For example, we can set a specific time and date to discuss how to best support religious (or other spiritual) organizations with our spouse, then automate financial commitments via payroll deductions or automatic withdrawals from a bank account. Or, when we're focused on God and our spiritual lives, we can create a commitment that is difficult to walk back from, such as organizing a Bible study group with friends.

Being Timely

In our daily lives there are countless tasks demanding our attention that urgently seek to be acted upon. It's no wonder sometimes that our spiritual lives can get pushed aside for things that feel more immediate. Our minds seem to prioritize the immediate and the urgent. So, *what should we do to keep our spiritual needs from being endlessly postponed or pushed aside?*

A range of strategies that build on our mind's bias towards urgent things in the present can make spiritual action timelier.

Consider the immediate costs and benefits

Many of the good things we want to do in our lives are about future rewards or future costs. For example, eating healthily is often seen as a way to live longer and avoid issues like heart disease. Unfortunately, both heart disease and a longer life are very far away, so they are simply not as urgent as one's immediate needs and considerations.

To give a spiritual example, giving to money to the poor is something that the Bible tells us to do, and we will have rewards in heaven. Our time in heaven, we hope, will not be very soon however. And while we should do good for good's sake, the added benefit of reward in heaven isn't quite as powerful as it might be. So what can we do? We can find ways to focus on current or immediate costs and benefits. For giving to the poor, it might be intentionally thinking about how we are helping that person right now — to buy food, to take care of their family, to enjoy their Thanksgiving meal, to provide them with a coat for the cold weather, and so forth.

Attach to an otherwise urgent or timely action

One of the cleverer techniques to make something urgent is *temptation bundling,* which we briefly discussed in Chapter 1. That's when there's something that you really *like* to do (and has an inherent urgency for you), and something that you really *should* do but you find you just don't have the same motivation for, and so you only allow yourself to do the "want to do" activity during or after the "should do" activity.[77]

The original research study looked at people who struggled to go to the gym — something they knew they should do, but sometimes faltered in the follow-through. The researchers paired this "should" activity with a "want" activity of listening to an entertaining audiobook. They gave participants free access to an audiobook, but only at the gym. They found that by combining these two activities, the incidence of the "should" activity increased remarkably.

Create or find specific deadlines for action

"Good" actions like exercising, saving for the future, and yes, reading the Bible, often have no inherent urgency: they are things we know we should do, but keep putting off and putting off because there are more urgent things we need to do. One way to address this is to *create* urgency. Marketers use this technique frequently, with "limited time offers" that

---

[77] Milkman et al. (2013)

are concocted out of thin air, but still motivate us to action so we don't lose the chance.

We can make something timely by setting a specific deadline for ourselves, even if that deadline is entirely arbitrary. In the 401(k) retirement space, HR often sets a specific deadline by which people have to decide on their contributions. With automated computer systems that can change someone's contributions at any time, there really shouldn't be any deadline; one day is the same as another to the software. However, one day being just the same as another to a person means there's no real urgency. And so this deadline, in the form of an "open enrollment" period, focuses people's attention and helps them take action where otherwise they would procrastinate.

In our spiritual lives, let's say we wanted to follow the biblical injunction to "forgive our enemies." Is there really a difference whether we forgive them today or tomorrow? Not really, especially if we've held a grudge for a long time. But if we intentionally *set a deadline and plan for that date,* we can create a sense of urgency around that date. That's especially true if we also make it social: arrange to meet with the person and have to spend time with them on the day of the self-imposed deadline. That fear of coming face-to-face and not having followed through on the commitment to forgive them in our hearts can make it all the more urgent to do so beforehand.

We can also look for and embrace external deadlines. On a personal note, a few years ago, I enrolled in part-time classes at the Earlham School of Research. One reason I took the classes was that I couldn't go whenever I wanted; their normal classroom programs require someone to take classes full time, often by moving in person to Richmond, Indiana. Well, I couldn't do that. But they had a "winter intensive" program for two weeks in January, where you come, take a class for that time, and go back home. So I talked with my wife, and we could feel the urgency of that deadline: I wanted to go to Earlham in person, and in December I could feel the opportunity slipping away. So I jumped on it. And once I signed up for the course, I created new deadlines for myself to read the background material beforehand, etc.

Things that I otherwise could have done *any* day suddenly had to be done, and by a *specific* day.

Use prospective hindsight

> Imagine a day in the far future, when you're approaching the end of this life. You haven't been perfect, but you've done your best to live a good and godly life. You're at peace with your family and neighbors, you're focused on the Spirit, and you await the next stage. Now, look back on what choices you've made with your time and energy: *What brought you to this point? What did you do right, what did you avoid, and how did you keep yourself on track?*
>
> Now imagine a very different future. One in which you're spiritually empty. Alienated from your faith and from God. Look back from that future day, on the choices that made that happen: *What did you do wrong? What temptations did you give into? How did you veer off-track, and why couldn't you recover?*

This technique, known as **prospective hindsight**, is one used in retirement planning to make the future more vivid and real, and to help people take timely action to improve that future.[78] In a retirement context, the exercise is of course slightly different, but the lesson is the same: we can focus our minds on the importance of our actions now by running through scenarios about the future and considering how our actions now affect those future outcomes. In each case, the goal is to help us use our precious time and energy a little more wisely, and a little more in line with what we truly care about.

What Should We Change?

Often, when people struggle to turn intention into action, they do one of two things: either they give up, or they beat themselves up. In giving

---

[78] See Mitchell et al. (1989) for initial work in the field; see Benartzi (2015) for an application to retirement.

up, they may believe that their past failure simply means that they're not "good enough" or "strong enough" to do what's needed. In beating themselves up, many seem to believe that if they simply muster their willpower and "man up" (or "woman up"), they can push through and do what's needed.

You may have noticed that across the range of techniques we've talked about so far, the approach is really quite different. Behavioral scientists see the process of changing behavior to make it more in line with one's intentions from a unique perspective: *they look for the specific obstacle, which usually occurs in the environment or something about the action the person wants trying to take, and they change that.* Behavioral scientists, in most cases, don't try to fundamentally change the person or to change human nature. We don't try to remove a person's biases or limitations: instead we try to work around them or with them.

To understand this approach, think about the following story.

> *Someone is stranded in the desert and is dying of thirst. There's a small spring at the base of a valley they are traveling in. After trying and trying to reach it, they collapse on the desert floor unable to move any further. You happen upon them and see their plight. What do you do?*
>
> *Do you yell at them for being weak?*
>
> *Do you tell them how important it is for them to keep trying to reach the water and how it's necessary for them to survive?*

Doing either of those things would be foolish. The person isn't weak; they're just dying of thirst! The person already knows how important it is to get water, so it's not motivation that's lacking.

Instead, if you wish to help the person, you do one of two things. Either you walk down to the spring yourself, gather some water, bring it up to the person and encourage them to open their eyes, open their mouth, and drink that water you've brought. Or, if you have no way to bring the water to them, you scout out the route, find the easiest path, remove obstacles along the way, and help them get to the spring.

The first, foolish, path (yelling) assumes that people fail because of something that is wrong inside of them. The second foolish path (educating) assumes that people fail because they don't know how important it is to succeed. In both cases, the assumption is that in order for the person to be successful, they have to be either educated or fundamentally changed.

So, what does it mean to instead gather water and ask the person to drink? It means this: we *find a different action that is easier for the person to take* — in this case, mustering the strength to drink the water you've brought, rather than to walk to the stream.

So, what does it mean to remove obstacles along the way? It means *change the environment so that it is easier to take the action*.

Don't get me wrong. It's not that working to somehow change the person — through education, training, reframing how people think, etc. — is necessarily the wrong thing to do. It's simply that we shouldn't limit ourselves to those techniques, and there are often more effective routes to be used, individually or in tandem with education and such techniques.

From these three different approaches, a general lesson from the behavioral research emerges: to change our behavior in the face of obstacles, *it is often easier to change the action (what we're trying to do) or to the change the environment to make action easier, than it is to change human nature.*

We can and should design our environments and behavior to support spiritual growth and practice.

The Kitchen Sink

In the past four chapters, we've talked about four ways in which to support intentional action, like a spiritual discipline or practice, using the acronym EAST: making it Easier, grabbing Attention and removing distractions, creating supporting Social connections, and making it Timely. We've talked about them as if they were separate ideas, which of course they aren't. It's just useful to introduce the ideas that way, to make them clearer. In real life though, we often use many techniques, all at once, to aid action. We "throw the kitchen sink at it," as it were.

As I look to follow a particular discipline, I don't care about whether a particular social technique or an approach to make it timely might be most supportive. I just want to spend more time focusing my thoughts on spiritual life. And even when I'm not consciously thinking about it, multiple forces weave together in my environment to focus or distract me from that world.

My aunt was a retired Presbyterian minister, and when she needed to live in a community with nursing and support staff in residence, she chose to live in a community with many other Presbyterian ministers. While she didn't make that choice because of the presence of the other ministers, as a behavioral scientist, I've thought about the likely effects of such a community on its members. They are fascinating and diverse. The community would create descriptive norms of religious practice, a community of discussion, physical reminders of spiritual life that ministers would display, and fewer frictions to spiritual conversation.

However, while these diverse factors are interesting and show how multiple behavioral mechanisms can be at work at once, that's not how we should think about applying behavioral lessons in our lives. Instead, *what really matters is whether our environment supports the life we want to live.* (For my aunt, her community did that.)

If we're able to live our faith and our environment helps us, excellent. If it doesn't, then the tools of spiritual design may help, whether in isolation or brought together in a broader package. We don't pick behavioral techniques, like prospective hindsight or descriptive norms, because they are interesting; we throw them, individually or as part of the proverbial kitchen sink, at a problem in our lives, to help start a spiritual practice or discipline.

And, after we've started a practice — perhaps with the aid of one or many of the techniques discussed here — the next challenge is to *sustain* that practice over time. That's what we'll turn to in the next chapter.

A Few Lessons

We've all faced times when we lack urgency or procrastinate. Sometimes we procrastinate with remorse, sometimes with resignation, and sometimes with glee. But when we actually want to act — to let our lives speak — we need better tools to overcome this challenge. This chapter sought to provide some techniques you can try in your own life:

- Recognize that most spiritual practices are *important but inherently not urgent*: there's no spiritual calculator that says that if you take the time to pray today instead of waiting until tomorrow, you'll be "rewarded" (and indeed for most Christians that's not how things should work).
- Most people are inherently *focused on the present*: it's more real, it's more vivid, and it screams for action more than the uncertain future.
- So, we can *think about the good we can do now*: for example, the help we can offer someone in need right now (immediate, vivid), in addition to the long-term value of loving our neighbors (vague and in the future). Or the sense of peace that quiet meditation gives us, in addition to how it reinforces our long-term spiritual journey.
- We can *tie into other, more urgent or desirable, actions*: like scheduling a half-hour of prayer time right before we allow ourselves to watch the latest episode on our favorite series on Netflix.
- We can *create deadlines or a constrained schedule for ourselves*, like enrolling in a religious study program that meets at certain times and dates — turning the vague sense of "I should read more" to "I need to have read this by Thursday at seven o'clock when the group meets."
- *Look back with insight, by imagining ourselves in a good and happy future, and also in one that isn't* — and thinking about what we did that brought us there in our spiritual lives. This helps

make the meaning of our choices more real and immediate to us.

And finally, *throw the kitchen sink at it!* While each of the last four chapters have talked about a specific behavioral factor and set of techniques (ease, social dynamics, etc.) to make them easier to present, real life doesn't need those neat boxes. Feel free to weave in any technique, individually or in combination, to help you on your path. In the end what matters is the value it brings to your spiritual practice, not which chapter or header it falls under.

7
SUSTAINING A PRACTICE

Our virtues are habits as much as our vices.
— *William James,* Talks to Teachers on Psychology

Shortly after I joined the choir at our church, I had a particularly powerful experience, one which warms me whenever I think about it. In our choir, some of the songs we sing are different each Sunday and some are repeated every time: they're part of the liturgy. Before joining this church however, I'd come from a Quaker tradition without singing or liturgy — so everything the choir does is new. After a few weeks, though, one of the simple songs we sing began to resonate with me: the Agnus Dei. It starts like this:

Lamb of God, you take away the sins of the world; have mercy on us.

Whenever I hear, and especially whenever I sing, that song, I am flooded with a beautiful feeling of peace. I feel it even as I write these words. And so, after a few weeks in the choir, I started to sing it softly to myself throughout the day. Sometimes I'd sing it out loud, but usually I'd sing it softly and slowly inside my head — feeling the rhythm of the words and of the song.

In time, I started to hear the song in my head even without thinking about it. I'd be thinking or doing something else and I'd turn inward and discover the song continuing as an unnoticed and unceasing foundation to my thoughts. I'd return to that song when I felt the desire to contemplate God — and I could feel its continuing rhythm pulling me in as well. My wife even tells me that I would sing it softly while I slept. (I had no idea I was doing that.)

Over time, the centrality of the Agnus Dei in my prayer and meditations lessened. Other songs and other non-sung prayers have filled that space, and brought other ways to focus on God. But the memory of that time months ago is quite special.

The inward rooting of the song is not something I truly controlled. But for my part, I was open to it, and I supported it with repeated meditative practice. If I had watched Hollywood movies instead of opening myself to prayerful song, would I have nevertheless had that special experience? Perhaps, yes; that's part of God's mystery. But *routines and habits of prayer like this can helps us listen and help us notice those special moments when they are there to be found.*

Each person has their own routines that help them focus on God; their importance and value for us is individual and personal, but we can learn from how they work in our lives, and how to help cultivate them, even if in the end we do not control them.

Moving From a Single Moment to Many

The tools described thus far have been about taking action in the moment, according to one's faith — for example, designing your environment to make it more likely that you'll pray or give to the poor. But when we have the desire to change, and live according to our principles, we want more than a momentary change. We want to lock in that change for the long term, and become a more faithful person over time.

For long-term change, we have an additional set of tools, each of which works in different ways. The tool that is most familiar to us is habit:

our mind's ability to execute repeated, frictionless routines. Other tools we can use that drive long-term change are our *internal self-narrative, our knowledge and skills,* and *feedback loops,* which help us adjust course when we're off-track. Let's start with habits.

Building New Habits

Imagine that you've started a new practice, like reading a devotional each morning. It's fulfilling and energizing, but you're afraid that you'll become distracted in your hectic daily life, and let this new practice slip away unnoticed. Habits can help maintain the new practice. They allow us to get distracted, as we are all prone to do, and yet still continue on a course of action. Habits are one of our minds' most powerful tools, and there is a growing body of research on them.[79]

The way habits work is by our minds learning, over time, to trigger a particular routine whenever we're in a particular context or environment. So we don't even think about it! Indeed, that's why habits are so valuable: they allow us to think about one thing, and do another. For example, imagine you were walking down the street in the morning next to your office when you smelled coffee and automatically walked into the coffee shop and ordered. You might have been thinking about something completely different, lost in your head, but your coffee buying habit was ready and waiting, triggered by the right context for action.

How habits are created

Habits are formed through repetition. Do something often enough, in a stable context, and you'll have the tendency to do it again when you're in that context. For example, when you are driving and the light turns red ahead of you, you press the brake. When you were first learning to drive, recognizing a red light, remembering which foot to press, and then doing it without causing the car to lurch to a stop — that was

---

[79] See Wood and Rünger (2016) for an overview.

difficult! But, in time, it became effortless and didn't require conscious thought. It became a habit. When you see a red light now, you naturally press the brake, even if you're talking with someone or listening to the radio.

Habits often form because of intentional goals: we want to accomplish something, and so we undertake a particular routine repeatedly in order to achieve that goal. Our minds diligently watch for the things we do regularly and put them on autopilot automatically for us. Good habits make it easier for us to reach our goals.[80]

While repetition alone is enough to form a habit, when you receive a reward of some type at the end of the behavior, the habit forms more quickly. For example, if you try to build a regular routine of spiritual meditation, you might feel a tremendous sense of calm and wholeness because of it. That reward reinforces the link between the cue and the routine, building the habit.

While intentions, goals and rewards are often part of *forming* a habit, they are surprisingly irrelevant in the execution of a habit, once formed. In fact, one of the definitions of a habit that researchers use is "a repeated behavior that is insensitive to outcomes" — as people do them regardless of whether the outcome is good or bad. Habits truly are an automatic response, and are triggered without conscious thought, seeking of a reward, or even an intentional purpose.[81]

A key lesson from the research, and for this book, is that habitual behaviors make up a significant part of our daily lives. Our habits help define who we are — *and we can help define our habits*. We can create habits through intentional repetition of a routine, starting after a clear cue or trigger in a stable environment, and often with a reward at the end.

How long do habits take to form?

The honest answer is that there isn't a general rule, and that solid research is still limited. Common wisdom about it taking "twenty-one

---

[80] Wood (2017)

[81] Wood and Rünger (2016)

days to build a habit" is simply false. Sorry. That particular bit of "wisdom" is an overgeneralization of a plastic surgeon's observation in a book from 1960;[82] it has no scientific backing. What appears to matter is *frequency, stability, and reward*.[83]

The more *frequently* the person does the routine, the faster the habit forms. The more *stable and clear the context* (the cue for the habit), and the less thought required to decide if one should perform the habit, the faster the habit forms. And, as mentioned above, when there is *a reward* — especially one that occurs after a time delay (a.k.a. an interval reward schedule) — habits form more quickly.

People and circumstances vary widely, though. In one of the few studies of habit formation over time, researchers found that the time required for habits to form for a set of daily tasks ranged from 18 days to a whopping 254 days.[84] Moreover, habits aren't on/off switches. Behaviors can become increasingly automatic over time — one can be partially on auto-pilot, but not have a full habit until it is repeated often enough to reinforce it and create a fully automatic habit.[85]

One way to find stability and frequency is to build on an existing habit, such as brushing one's teeth in the morning or taking a shower. For example, right before, or right after, brushing one's teeth (or taking that shower), do X. People generally brush their teeth (and shower) at the same time each day, whether that is a point in time on the clock, or a point in time as part of a going-to-bed-routine. They also tend to do it in the same place. That potentially sets up a good situation for new habits to form that are linked to it: like learning to take your vitamins by doing so before brushing your teeth. This approach of linking new

---

[82] See Gardner (2012) for the story

[83] See Wood (2019) for a detailed description of how habits work by one of the leading researchers in the world. See Wood and Neal (2009) for a shorter summary. See Duhigg (2012) for an accessible layperson book.

[84] E.g., Lally et al. (2010)

[85] Ibid.

Spiritual Design

habits to prior ones has been popularized by Stanford social psychologist BJ Fogg.[86]

The downsides of habits

Habits have two significant weaknesses. First, once they are formed, habits are automatic and mindless. With many habits (like pressing the brake when seeing a red light), we may not even realize that we've done it (neither the cue of the red light nor the routine of pressing on the brake are conscious). Now, imagine *a habit of prayer* in which you automatically knelt, said some words out loud or in your head, got up and continued about your day. To do all of that nonconsciously would be empty; for Christians, it would be a horrific example of following the Law devoid of the Spirit, or an act without faith.

So, if executing a habit is spiritually empty, why do we talk about habits at all? It's because habits can play another role in one's spiritual life: by setting the stage for intentional action. For example, in my life I have the habit of pausing before each meal, bowing my head and clasping my hands — i.e., preparing to pray. *What I actually do in prayer isn't part of the habit at all.* I sit and listen for God's word in my life. Sometimes that takes a few seconds, and sometimes it takes minutes. If I voice any words within, they are unscripted, and based on the moment. The habit of preparing to pray is immensely useful for me though, because it regularly and automatically sets up the situation to engage in non-automatic, intentional prayer. Without that habit, I would be consciously trying to remember to pray when I'm distracted by other things, or in a hurry to eat.

My example isn't meant to say that all scripted prayers are empty; rather, that the parts that are habitual (the words themselves, for example) become nonconscious and non-intentional. *What we do in that potential moment of reflection is what really matters.* We've all been in situations where those moments of potential reflection quickly turn into a review of the day's to-do list. And at other times, they allow us to gain a

---

[86] See, for example, TinyHabits.com and Fogg (forthcoming).

profound sense of peace and connection with God. It's not the habitual act itself that matters; it's *what we do with it, through our faith.*

Similarly, we can use habits to set the stage for later faithful action. For example, if we have a deeply engrained habit of getting in the car and driving to church on Sunday that is nonconscious and non-intentional, it says nothing about us as a spiritual person (it's spiritually neutral). But, once there, we can open ourselves up to God's word and the community of believers, and grow spiritually. Bringing the Bible with us when we travel can be a habit that itself is neutral, but very is useful because it makes it easier for us to then actually read it.

Finally, while acting on an existing habit — such as going to church — is spiritually neutral, building the habit is not. By intentionally resisting the desire to sleep in, and by facing difficult spiritual questions and embracing spiritual growth, we do a conscious and effortful thing. *Building habits can help us grow and change ourselves and our routines to better express our faith.* In other words, building habits is one of the ways in which we demonstrate our faith by our deeds. We must, however, be vigilant that once the habit is formed, our faith still accompanies and enlivens those deeds.

In addition to being mindless, habits have a second weakness: they are fragile with respect to their environment. If the smell of coffee weren't there, or you were out traveling and didn't have your toothbrush, the habit linked to that cue wouldn't trigger at all. It's for that reason that one can avoid *bad* habits by changing the environment in which they occur. But for good habits, that fragility is a problem. And so we look at how to make the environment more stable — for habits, and for other more conscious changes in behavior.

The Importance of Making Structural Changes to Our Environment

If placing the Bible by the bedside helps us remember to read at night, why wouldn't it continue to do so in the future? Well, it certainly might. And you might repeat it often enough that a habit forms and you don't even need to think about picking it up until you already have it open and are finding your page. What happens if you pick up the Bible one day and bring it to the living room, though? When you head to bed, it isn't there to remind you. To counteract this, you can build in changes to your environment that are more durable.

The idea is to *lock in and 'institutionalize' aspects of your environment that help you*. Whatever you found in Chapters 3-6 to support a one-time action, you can plan ahead to make sure that it is still available for you in the future. For example, if a calendar reminder helps you block off your schedule for a gathering of your religious community on Wednesday night, you can make it a reoccurring and indefinite reminder. For some people, physical tattoos serve as an even more permanent and reoccurring reminder of a spiritual commitment.

Other structural changes include moving houses or changing how one gets to work. In my life, taking the metro is a central part of my spiritual routine: it gives me a consistent time to focus on spiritual matters when I don't have to focus on driving, taking care of the kids or working. For others, intentionally moving closer to a religious community can make it much easier to lock in new and still delicate behaviors of sharing time with the community.

Certain types of spiritual actions can be locked in even further by *automating them*. For example, let's say you contribute to your church by putting some money in the plate. You want to contribute regularly, but sometimes you just don't have cash; this is actually a frequent problem for me... and an embarrassment. You can take the desire to contribute and lock it in, by automating it. Most churches (and I imagine, most synagogues, mosques and ashrams) allow members to set up automatic contributions from their checking account straight to the church, on a regular schedule.

If you recall the metaphor from Chapter 6 of a person dying of thirst, think about what is most effective at changing behavior in general. Instead of trying to change the person (adding motivation to seek out the water they crave), we change the environment (make it easier to find water), or, ideally, change the action (bring water to them, so they just need to drink it). Automating a tithe is an example of changing the action.

Recasting Our Self-Narrative

A *self-narrative* is the story we tell ourselves about who we are, based on our past experiences and our understanding of our future path. For example, our self-narrative may say that despite everything we've tried, we can't find the strength to live in the manner we want. Or it might say that despite everything we've tried, we can't find the strength to live in the manner we want *without the grace of God*. Both of these self-narratives might "fit" the facts of our past experience, but they'd lead us in very different directions in the future.

The phrase "self-narrative" has been popularized by Tim Wilson, a social psychologist at the University of Virginia, with his book *Redirect*.[87] Wilson studied how to help people 'edit' the self-narrative by reinterpreting their prior experiences. For example, by helping people reinterpret past problems not as signs of permanent failure, but rather as the normal steps one takes to learn and to do better in the future.

Our self-narrative of the past shapes what we will try in the future, and what we will consider beyond our reach. In other words, your self-narrative can rob you of hope, or it can help provide it. And so if you feel that sense of hopelessness — that who you now are will not allow you to act in a more spiritual manner — that belief can become true and self-reinforcing. But it needn't be that way. You can prove that self-narrative wrong by succeeding — and you can directly attack that self-defeating self-narrative by rewriting the story.

Now of course we see examples of optimistic (and pessimistic) self-narratives in various areas of religion. At one end of the spectrum, there is *prosperity theology*, which teaches us that naming and describing ourselves in the way that we want to be, rather than in the way we are, helps bring about that new state. But in smaller ways, we can reinforce

---

[87] Wilson (2011)

the good of our past, and the times in which we have done right (whether that be turning our minds to God, giving to the poor, or loving our neighbors). We can see ourselves as people who have the strength and prior history to do these things again, even if we haven't been doing them as often as we'd like. *By focusing on examples of that better self, we can help make it more prominent, more real in our lives.*

Feedback Loops

One of the enduring lessons of both behavioral science and the Christian faith, along with many others, is that we simply aren't perfect. We make mistakes, both in our daily lives and in our attempts to improve them.

We can and should look for ways to do better: to fill ourselves with prayer, to plan thoughtfully ahead, etc. But we'll still fall short. And so we try again, and try again. To make those repeated attempts valuable and get us closer to the person we want to be, we need more than perseverance: *we need to know that we've erred.* We need to know how far we've missed the mark, so that we can improve next time.

That is the heart of the feedback loop: to try, to fall short, to know that we've fallen short (and how), and to try again in a new way. It might seem trivial. We know when we've failed to accomplish something, right? Actually, though, we're really good at telling stories to ourselves — and warping reality to make us feel better.

For example, let's say you are trying to spend less evening time at the office and get home earlier. You want to spend meaningful time with your family, such as talking about all of the things we are thankful for at dinnertime. You've done some things to make it EAST (easier, attention-grabbing, social, timely) to do so. You feel good about it. But, over time, are you likely to continue and actually improve?

Without help, you're unlikely to have a clear and accurate picture of whether things are getting better or not. Sadly, this occurs for many reasons documented by behavioral scientists. Here are two of them:

- **Availability heuristic.** Our minds judge the frequency of past events by how easily they come to mind, which is often shaped by how vivid or *unusual* an event is. So things that "come to mind" easily are considered more frequent than they might actually be.[88] For example, you'd be likely to selectively remember the few times you did come home early, and how wonderful that was, but not notice that, on average, you're actually spending the same amount of time at work.

- **Confirmation bias.** As we seek to answer questions (like, "Am I doing better?") and gather information, we selectively pay attention to, interpret, and remember information that *supports our pre-existing beliefs*.[89] In the case of getting home early, you might unintentionally pay more attention to praise from your spouse on the few days you do return early, rather than the negative comments.

There is a way to overcome these problems, though — *with an external reference point that tracks your progress and helps you learn*. In other words, a feedback loop. In this situation, a feedback loop for spending time with family would require an external way of keeping track of one's successes (and failures) — for example, a simple paper log or an electronic note or app on the phone in which you tracked the time you got home (and whether you spent the time there as you'd hoped) every day. After trying to do better for a week or so, you'd check the log to see whether you actually succeeded — regardless of what your memory says. Having a feedback loop like this can keep you on track as you change behavior over time, and is especially helpful when you aren't living up to your goals. In those cases, sometimes we need a bit of extra help to try again.

---

[88] See Tversky and Kahneman (1973) on the availability heuristic.
[89] See Nickerson (1998) on confirmation bias.

Trying Again

We're imperfect beings, and sometimes we fail to act with faith despite our best efforts. The techniques above can be used to prepare for a moment of action, and make it more likely one will act according to one's faith. However, if you are struggling to act with faith, and feel that you've tried many times before but failed, there's a specific technique that you can use that might help: the fresh start.

Fresh starts are special times in our lives when we feel a new opportunity to change something about ourselves. Research by Hengchen Dai, Katherine Milkman, and Jason Riis at the University of Pennsylvania about fresh starts[90] finds this: *People are disproportionally likely to make major life commitments during times of transition.* For example, these scientists looked at how people are more likely to make commitments like exercising more or eating better over New Year's (New Year's resolutions), birthdays and marital changes. The behavioral logic is this: When we've struggled to do something in the past, fresh starts give us a reason to hope things will be different this time around. We mentally separate out our experiences from before the fresh start, and label them as irrelevant or outdated ("That was *last year!*"). The time after the fresh start has a newness free of our historical baggage that lets us try something different, or recommit ourselves to a prior goal we've failed to achieve.

If there's something we've tried to do in the past but struggled with, like attend church regularly, then in these special fresh start moments we can have renewed vigor and a sense of hope that we otherwise would not have, given our prior experience. For example, some people make resolutions on New Year's Day to be with the kids more often or go to the gym — or in the spiritual realm, read the Bible regularly. Even though one may have *tried* to do all of these things in the past, the fact that it's a New Year's resolution makes it feel special and allows

---

[90] Dai et al. (2014)

us to put our past experiences in a separate historical category that doesn't doom us to repeating those mistakes again.

Personally, the times when I've made the biggest steps in my spiritual life have been at moments that were otherwise "special" or signified a new beginning or stage. For example, when my wife was pregnant with our first child, and when we moved to Chicago, we made commitments to ourselves to bring religion more into our lives. For Alexia and I, we felt a sense of newness and an opening up of possibility. We felt that these were times we *should* make a change. And we did.

I'm not aware of any behavioral research specifically on the act of confession, but it would seem to provide a similar fresh start, allowing people to redouble their energy towards spiritual practice even when they've made mistakes in the past. And, for Christians who are reborn in Christ, the acceptance of Jesus in one's life is the ultimate fresh start.

In your own life, you can think about special days like birthdays, holidays, job changes or spiritual milestones that create a natural break between the past and the future. *The future can be different, if you make it so.*

Entering a Virtuous Cycle

Spiritual practice and faith can reinforce one another over time, supporting a personal transformation that has both outward and inward dimensions. The more we act with faith, the more opportunities we have to interact with God's Word and his Church. The more we learn to open ourselves to God, the more we naturally seek to act in accordance with our faith. It can be a virtuous cycle.

Here's a simple example. Going to church is part of being faithful to God's teachings. It can also support further growth in one's faith. Entering the building doesn't guarantee that one actually listens and acts on God's messages. But it's much easier to learn about God, feel the support of a community of believers, and perhaps deepen one's faith, than if one goes to see movies on Sunday morning or sits at home reading the newspaper.

We can see this virtuous cycle elsewhere, in the most basic of actions. We might choose to place the Scriptures on our bedside table, as a behaviorally-inspired technique, because we believe that the Word of God is sacred, and that the Bible is often the best place to start to better understand God and follow His will. Placing the Bible within reach can also have an indirect effect on our faith itself: when we read the Scriptures, we gain insight about God, and more subtly, about ourselves and our relationship to God. And, as our faith in God strengthens and becomes richer, so too may our desire to act according to that faith — showing a virtuous cycle once again.

A Few Lessons

For most of my life, my religious experience was that of Quakerism, and I never had a strong foundation of religious habits. Quakers don't pray the rosary; we don't state a creed. We don't even have any pre-planned or regular structure for our Meetings for Worship (a.k.a. church). That has changed recently, as Alexia and I have become a part of a deeply spiritual and liturgical Lutheran Church near us. It's a vastly different experience, and one that has helped illuminate the power of routine and habit.

Many Christians, especially ex-Catholics, talk about how religious rituals have become empty for them, devoid of the very spirit they were meant to convey. But, as I start to adopt these rituals myself and think about how habits work in the mind, I think that the formatting of habits is quite different than their habitual execution. And that when we understand both phases of habits, we can shape them to play a more positive role.

There are a few lessons from this chapter that might be useful in your own life, as you seek to build meaningful routines and otherwise build upon a new spiritual practice to make it an integral part of your life:

- Habits are, by definition, actions without thought. They are neutral on their own. But we should not shun them. *If used wisely, habits can create the mental and spiritual space for deep and meaningful contemplation.* They can help us lock in changes in our routine over time.

- Another way to support long-term changes in our lives is to *make structural changes in our environment:* for example, with a set of cues to act with faith that we'll always see. An extreme example would be to move houses to live next to church: you'll see the church every morning, and be reminded of it! A less extreme example would be to setup a reoccurring calendar item to take the time to pray, to establish a regular spiritual discussion group, or to setup automatic deductions from a bank account for charitable contributions.

- Especially when we've struggled in the past to act with faith, *the narrative we tell ourselves about our past actions matters,* including what that narrative means about who we are as a person. If we get discouraged, we can look at that narrative and ask if it's really true and the only way to view our past. As a behavioral scientist, a key part of my self-narrative is that we're all imperfect and subject to our environment. That narrative doesn't discourage me though; rather, it helps me understand that stumbles are natural and to be expected. *We can seek to do better despite everything we've failed at in the past.*

- Since we all stumble on our path, it's important to be able to start anew: to take a step back, forgive ourselves, and try again with openness and energy. *Fresh starts* are one route to do that. If you are discouraged, look for the natural moments of change in your life — the start of the year, a birthday, a move to a new city, etc. — and use that time to press the reset button.

- Along the path to long-term change in our lives, it helps to set up a *feedback loop* to see how we're doing along the way. This might be with a regular check-in, reviewing what you wanted to do and what actually happened. It might be a trusted friend who gives you honest and critical feedback, for example, of how you are treating others around you, and whether you are truly living up to your desire to "love your neighbors."

Each of these techniques are about encouraging and supporting actions you want to make an ongoing part of your life. But there are also actions you might want to *stop* doing, that are contrary to your faith and how you want to live. Perhaps you've been judging those around you harshly, or speaking to a particular coworker with malice, not remembering and honoring our common relationship in God. You don't like doing it, but you do it anyway. *How might you overcome these habits and ways of acting that are contrary to your faith?* Let's look at that next.

8
AVOIDING BAD HABITS AND OTHER CHALLENGES

Your task is not to seek for love, but merely to seek and find all the barriers within yourself that you have built against it.

— *Rumi*

Within Christianity and many other faiths there are both prescriptions on what *to do* in one's daily life and injunctions on what *not* to do. And indeed, there are techniques we can use to *take* action, and different ones we can use to *stop* something.

To start a spiritual practice, there's almost always an element of conscious thinking and preparation. That's what we've talked about thus far: tools to help us act according to our faith, especially in starting a new discipline or spiritual practice. As you've seen, there are many lessons we can draw upon, since the bulk of research in behavioral science is on helping people take specific actions (though not spiritual action per se).

To stop doing something, we often require another set of tools because many of the actions we take in daily life simply aren't conscious. At least, they are not conscious in the sense we normally imagine: with a declarative decision to take a particular action. Instead, our actions, both good and bad, are often driven by our habits and our intuitions, each of which are themselves often shaped by the specific details of one's environment. We should understand that process, and our responses, to help change them.

Our Biggest Obstacle

Many times, *we are our own biggest obstacle to living according to our faith*. We habitually drink, even though we know how badly it may make us behave. We give in to temptation and binge-watch TV instead of devoting time to contemplative reading. Or we simply make bad choices that we'll later regret.

Researchers study these challenges as distinct but related problems.[91] Simply put, habits (automatic responses) work differently than temptations (emotionally charged responses), and differently than bad choices (conscious, but narrow or short-term thinking).

Thus, the first question one should ask in trying to avoid an undesired behavior is this: *what type of behavior is it?* Namely, is it something we do routinely, again and again and again, as a nonconscious habit? Is it something that overcomes us with emotion or desire: for food, for drink, for sex, etc.? Is it a conscious and calm deliberative choice we make in a particular moment that we later regret and realize was wrong? Or is it something in-between — something we think lightly about it, or follow a gut feeling on whether to do or not, neither truly thoughtful and deliberative nor nonconscious?

Let's look at each of these questions and see how we can translate lessons from the research literature.

---

[91] E.g., Quinn et al. (2010)

Overcoming Habits

Have you ever had a long-term and pernicious habit — like binge-watching TV — and said to yourself, *Well, I just need to stop doing that!* as an attempt to change it? Chances are that probably didn't get you very far.

When we have a habit that gets in the way of our spiritual practice, we often try to end it immediately. However, since acting on habit isn't a conscious choice, a conscious choice to stop it can be remarkably ineffective. The mind automatically executes habits when it sees the right cue in the environment, and thus merely deciding *I don't want to do this anymore* isn't enough.

Constant vigilance

Habits are automatic, but they can be overridden by *exerting willpower*. You can think about habits as our *default* behavior in a particular circumstance. When we're not paying attention, or don't make a choice about what to do, prior habits direct our actions. When we are paying attention, and exert the effort, we can decide not to execute a habit and do something else instead. Sounds simple and straightforward, right? There are some very significant limitations.

First and foremost, we're usually not paying attention to every little thing we do in our daily lives. By some estimates, nearly half of our daily behavior is habitual.[92] During this time we can be thinking about other things — our families, our jobs, etc. To constantly watch over our behavior would mean not thinking about these other things. We'd be tying up our minds with *watching*, instead of *living*. It's exhausting to stay vigilant, going against our natural tendencies to shift attention (i.e., get distracted). It's also exhausting to stamp out the habit once we notice it; we have to intentionally and consciously inhibit what our minds would naturally do (the habit). And finally, habits are more likely to trigger in situations where we are distracted, stressed, or exhausted

---

[92] Wood et al. (2002)

— i.e., exactly situations in which we'd be less able to muster the strength to inhibit them. While this approach can be effective for short periods of time, thankfully there are better options.

Remove the cue

Our minds automatically execute a habit whenever we see the cue we've learned. *One of the ways to avoid bad habits is to avoid the cue.* If someone has the habit of stopping for a drink once they see the sign for the favorite bar, they can intentionally change their route home to avoid seeing the bar. If someone has the habit of binge-eating ice cream whenever they see it in the refrigerator in the evening, they can avoid buying the ice cream in the first place to remove the cue. Researchers find that habits (good and bad) are disrupted when people move[93] or otherwise have significant life changes that modify what they see and interact with on a daily basis. If you can identify what triggers a bad habit, you can intentionally avoid that trigger in the same way.

In my first book on behavioral science (Wendel 2013), I discussed a software package, Covenant Eyes, that helps people stop online habits, like viewing pornography, that they want to avoid. One of the ways in which this works is by blocking access to pornographic sites — i.e., removing the cue.[94] Covenant Eyes and numerous other such tools are grounded in a religious understanding of the destructive power of online addictions, though they can be useful for non-religious people as well, of course.

Moving beyond obvious habits like viewing online pornography, there is a significant limitation to applying this technique more generally: we need to know what the cue is![95] Habits can trigger off of odd, nonconscious things. For example, a smoker might be cued to smoke by the exit sign at work, having left by that door to smoke many times

---

[93] Bamberg (2006); Walker et al. (2015)

[94] See http://www.covenanteyes.com. In addition, Covenant Eyes creates an 'Accountability Report', which is an example of a social commitment we discussed in Chapter 4.

[95] Quinn et al. (2010)

in the past. It can be hard to disentangle and recognize that particular cue from amidst all the other possibilities: a particular time of day, a comment from a colleague, seeing others smoke, etc. To avoid the habit, the person would need to avoid seeing the exit sign, yet other life changes might not be as effective. If someone doesn't know their cue, they can't avoid a habit except through radical changes to their environment — disrupting many parts of their daily cues and routines in the hope that those changes include the actual cue for the bad habit.

Hijack the habit

We can target an existing bad habit and replace the routine with something more desirable. For example, don't try to stop the habit; instead, hijack it and use it for a different purpose. For example, imagine that you have the nasty habit of gossiping with friends about all of the bad things that others are doing (but never yourself, of course). The habit might be triggered by picking up the kids from school and interacting with friends there. You can't avoid picking up the kids, and you don't want to avoid your friends. But you can recognize the habit and intentionally build a new one on top of it. Use the same context, and same cue, to trigger something else, like sharing plans for the weekend.

Hijacking habits is a useful technique because habits, once formed, never truly go away in the mind. There is a hard-wiring of cue and routine: whenever you see the cue, you have the tendency to start the routine. While that wiring can't be removed, it can be overridden with a competing response; i.e., by forming a new hard-wired response that, over time, is more powerful than the original habit. So it uses the same cue and competes for control over your behavior. This technique is used as part of a habit reversal therapy to help people overcome the negative expressions of Tourette's Disorder.[96] In that community the technique is called *competing response training* — which nicely expresses what how the process works.

---

[96] Himle et al. (2006); see also Dean (2013)

The New York Times reporter Charles Duhigg, in his book the *Power of Habit*,[97] discusses how this technique often needs more than just a new routine to replace the old one: it requires faith that things will get better. Since the old habit never goes away, sometimes people will slip up and act on the old habit instead of the new one. This is especially true in the beginning of the replacement, as the old habit is stronger than the new one. But it can also happen years later. In the face of slipups, we need faith — religious or otherwise — that we can recover and continue on our path of breaking the old habit.

Mindfulness

Instead of conscious, constant vigilance against our habits, there is a less demanding way to overcome bad habits. A growing body of research shows how *mindfulness can help people avoid negative behaviors and feelings.*[98] Mindfulness, inspired by Buddhist spiritual practice, entails focusing one's attention on the current moment's experiences within the body and outside of it. A mindful state helps a person acknowledge the impulse to do something, but not act on that impulse. We learn to let the impulse simply pass. The result is a calm and intentional state that doesn't require an effortful fight to stop one's negative impulses.

Mindfulness training has been the subject of increasing research, and has been found to be effective in treating people with depression, anxiety, and stress disorders.[99] It is also the foundation for a recent approach in clinical psychology, Acceptance and Commitment Therapy,[100] and specific work on supporting good behavior (physical activity) in the face of bad habits like binge-drinking.[101] It is outside the scope of this book to provide a detailed description of how to engage

---

[97] Duhigg (2012)

[98] E.g., Baer (2003)

[99] E.g., Khoury et al. (2013)

[100] Hayes et al. (2011)

[101] Chatzisarantis and Haggar (2007)

in mindfulness, but Baer[102] provides a nice overview of the techniques used in clinical practice, and their empirical support.

What doesn't seem to work: punishment

Often, I've found that people beat themselves up if they do something they regret — whether it's overeating or ignoring people in need. They feel that if they were only "better" people, and had stronger willpower, they would do better. Willpower and the desire to change are certainly important across all types of behavior change. But when it comes to habits, adding self-punishment or anger when the change doesn't happen can be remarkably ineffective.

Once a habit is formed, by definition it is insensitive to outcomes. We don't engage a habit because we get something out of it. We just do it because it's hardwired in our brains through prior repetition. And thus punishing ourselves after we've slipped up and acted on our bad habit is unlikely to help. Instead, we need to give ourselves another chance — as we talked about in Chapter 5 — and look more thoughtfully at why a habit occurs and how we can avoid or work around it.

Addictions and habits

In thinking about habits, and what we can do about them, it is important to note that unfortunately these lessons may not apply to addiction. Addictions — drugs, alcohol, or behavioral — are different to and stronger than everyday habits. In addition to being automatic responses, they include a conscious and powerful craving that we feel deep within. They can require the help of a professional — and often strong faith in God — to overcome. If you have a habit you're trying to stop because it runs counter to your faith, take a moment to consider what type of habit it is. *Is it an addiction?* If so, you need more than spiritual design can offer – you might consider consulting a medical doctor or potentially a psychologist.

---

[102] Baer (2003)

If it is an action you've taken so many times that you don't even think about it anymore, then it is likely a habit. You can overcome it, just not in the obvious way of merely consciously telling yourself to stop.

In addition to habits, they are many behaviors we consciously decide to do, but regret. They aren't "habits" in the sense we mean here, but there are techniques we can use to overcome them — and which we'll talk about next.

Overcoming Temptations

The word *temptation* is complex, to say the least. Instead of diving into how the term is used in various contexts, let's talk about a specific way in which it's used in the research literature, and how research on overcoming it can be useful. Here we'll talk about *temptation* as *something that triggers a "hot state" in which a visceral emotion changes how we think and what we value.* Such emotions are ones we've all experienced, and include hunger, thirst, lust, anger and fear.

Temptations aren't all bad, of course. It's a generally a wise thing to respond to our temptation to eat and drink. But sometimes they are negative, and they derail our vision of how we want to live our lives.

What strategies are there to handle the "hot state" of a visceral emotion? First, it helps to understand what the visceral emotion does to us: it crowds out other thoughts and goals, making it difficult to focus on anything but answering the craving.

Second, we can look at what doesn't work. Thinking about and being vigilant against the temptation can be counterproductive. That's because, as researchers Quinn and others state in the context of food cravings:[103]

> The experience of craving develops through the elaborated thoughts and imagery that arise when people consciously focus

---

[103] See Quinn et al. (2010). Their comments build on Kavanagh et al. (2005).

on food and other desired appetitive stimuli… In these ways, a focus on the tempting stimulus exacerbates the power of the hot cues to active impulsive responses.

What *does* help? As with habits, we *can avoid the things that trigger our visceral emotions* (what researchers call *stimulus control*[104]). That can mean intentionally looking away, or avoiding the situation altogether. Similarly, we can *intentionally distract ourselves*: find other things to focus our attention on. Some of the initial research in this field comes from small children avoiding the temptation to eat a marshmallow.[105] The most successful children looked away or found ways to distract themselves with other activities. Not surprisingly, those who stared at the marshmallows tended to cave in to temptation.[106]

There are a variety of other ways that we can overcome temptations which are effective for exercising self-control overall. In the next section, we'll look at these broader set of tools. Before moving on from specific techniques for temptation though, one caveat: people often underestimate the power that visceral emotions will have over them[107] — which makes them harder to plan for and to avoid.

Exercising Self-Control Broadly

A common thread across fighting bad habits and temptations is that both entail self-control. *Self-control,* from the perspective of researchers in the field, is defined as *controlling thoughts, emotions, impulses, and*

---

[104] Quinn et al. (2010)

[105] Mischel et al. (1972)

[106] Since then, research on distraction has been conducted in diverse contexts from smoking cession to improving what people eat. Simple distraction isn't always the most effective; Versland and Rosenberg (2007), for example, found that a focused imagery exercise helped people more.

[107] Loewenstein (1996)

performances.[108] That's what's needed to monitor or hijack a habit and to distract ourselves from temptations — or more broadly, to avoid bad choices we know we'll regret.

In addition to automatic responses (like habits and triggering temptations), we use self-control when there's a gap between what we know we should do and what we'd actually do, or when there is a conflict between our goals, usually between our short-term goals and our long-term goals. Self-control tilts the balance of behavior from our short-term goals in favor of our long-term or "should" goals.

Self-control isn't just willpower: It's good habits, too

Researchers have found that people who are naturally good at self-control actually don't exert more effort to resist temptations and inhibit bad behavior in the moment[109]; on average, they aren't especially likely to inhibit bad habits. Instead, *they are better at forming good ones that keep them on track.*[110]

For example, people who build habits of eating healthy snacks and exercising regularly don't just inhibit the desire to sit on the couch and binge-eat: they have formed different routines. Students who establish good study habits are better able to stay on track with their studies because of what they are doing, and not just because they can resist habits of procrastination. Existing research has looked the benefits of

---

[108] Tangney et al. (2004). This discussion is about self-control overall, and not about the resource-depletion model of self-control. Over the last two decades, a popular model of how self-control works is that it is a limited resource that becomes exhausted and needs to be replenished — linked to glucose in the brain. That model has been drawn into question (Engber 2016). However, the core issue of self-control (and the understanding that people are sometimes able to control their thoughts and behavior, and sometimes are not) is an active area of research.

[109] E.g., Imhoff et al. (2014)

[110] Galla and Duckworth (2015)

setting up a regular, frequent schedule for studying, in a consistent location, for example.[111]

Spiritual design in reverse

> *By manipulating our circumstances to advantage, we are often able to minimize the in-the-moment experience of... struggle typically associated with exercising self-control.*
> — *Angela Duckworth*[112]

The environment we're in has a strong influence on our ability to resist temptations. For example, we're far more likely to over-drink when alcohol is present and available than when we need to go search for it. We're also more likely to need self-control to avoid over-drinking when others around us are doing so.[113]

Thus, one of the most effective ways to avoid giving in to these temptations is to selectively design our environment to minimize their power. Researchers call this **situational self-control** (i.e., *changing the situation*), and find that is often a more effective way to direct our behavior than effortfully trying to resist a temptation staring us in the face.[114]

By changing our environment, we can make ourselves less subject to peer pressure for example, making it harder for us to act in a way we'll regret. In other words, the same lessons we've discussed previously in this book for spiritual design to support good behaviors can and should be used in reverse to avoid bad ones.

What does this process of spiritual design in reverse look like? You can think about it in four steps:

---

[111] Ibid.
[112] Duckworth et al. (2016), Abstract
[113] See Hofmann et al. (2012)
[114] Duckworth et al. (2016)

1) **Double-check what type of behavior it is.** If it is habitual (nonconscious, without intention), check the techniques from the previous section on "Overcoming habits." If it is conscious and intentional, continue.
2) **Identify what supports the negative behavior.** Is it Easy, Attention-Grabbing, Social and Timely (urgent)? (See Chapter 3.)
3) Come up with strategies **to change the environment to create obstacles** — adding friction, removing cues, etc. — and to lock them in over time.
4) **Set up a feedback loop** to check in and see if you are being successful at stopping the behavior. We need feedback loops to stop behaviors just as we do when we're starting them, because we're just not that good at seeing averages and trends in our behavior over time.

Step 3, in particular, deserves extra attention: How do we *create* obstacles? We've spent a lot of time thus far talking about how to remove them. Well, here are some quick and simple ways to add obstacles to behaviors we want to avoid:

1) **Easy**: Figure out what's required to do the behavior you're trying to avoid — and *make it inaccessible.* This example isn't spiritually related, but I have the bad practice of checking the news many times a day. To make it harder to do so, I canceled my online news subscriptions and removed the apps from the phone. I can still look for the news by opening a browser, but that requires me to do more work: opening the browser, typing in the address, clicking on it, etc. That added *friction,* making it harder, slowing me down and giving me more opportunity to think twice about what I'm doing.
2) **Attention-Grabbing**: Just as the secret to getting attention is to put it in the line of sight, to avoid attention, *get the item out of sight.* For my wife and me, that's removing phones and computers from our room so we have fewer distractions from spending time with each other.

3) **Social**: How can we use our innate social sense to help us avoid bad behaviors? One way is to intentionally surround ourselves (at the moment of temptation or otherwise) with peers who *don't* engage in it, or who actively disapprove. Another way is to avoid being around friends who *do* engage in it and trigger us to do so as well.
4) **Timely**: How can we remove a sense of urgency from the parts of our lives that distract us from spiritual efforts? I'll admit I struggle with this one greatly, and I don't have an answer that has worked for me. I try to *increase the urgency of spiritual pursuits* — like setting a deadline for myself to write this book — *to crowd out other things*. That helps, but not as much as I'd like. Many people have found mindfulness practice helpful here: to remove the false urgency from life's many distractions.

Other techniques include:
- *Bucket your time* by reserving a certain amount of it each day for work, for family, for chores, and for your spiritual life. That is similar to Stephen Covey's "Sharpen the Saw" rule in *The 7 Habits of Highly Effective People*.
- *Do a goal-setting exercise* in which you compare how you are spending your time, versus how you want to spend it. Use that comparison to drop seemingly urgent but unimportant actions from your life.

A Few Lessons

The key to changing negative behaviors is to remember that *many of the actions we take in daily life simply aren't conscious*. For that reason, using willpower alone may not be enough to avoid or overcome them. Instead, if what we're fighting is an automated habit, we have a set of tools we can use based on how habits are wired in our heads. We can remove the cue that triggers our habit, we can override it with a new habit that uses the same cue, or we can engender a sense of mindfulness — which of course can be a valuable spiritual practice on its own. More broadly, for negative behaviors above and beyond habits, we can apply *spiritual design in reverse,* by making a behavior *less* easy, attention grabbing, social, and timely.

It's important to remember that these techniques don't necessarily apply to addictions, which are chemical dependencies in the brain. For that, we need another set of tools and supports, which are beyond the scope of this book. And indeed, there are a range of other areas in which we should be cautious and thoughtful about the limits of what spiritual design can offer; while spiritual design can be immensely helpful in our spiritual lives, it's not a panacea. And that's what we'll turn to next.

9
THE POWER, AND LIMITS, OF DESIGN

We must be ready to allow ourselves to be interrupted by God
— *Dietrich Bonhoeffer,* Life Together

I Can't Tell You What To Do

Throughout this book, I've tried to present the behavioral literature as offering a set of tools and lessons that people *might* find useful in their own lives. Why don't I just tell you what to do to be a more spiritual person or to lead a more a meaningful spiritual life? First, as a researcher, that would be dishonest: I mentioned earlier in the book that behavioral research specifically on spiritual practice doesn't seem to exist. The ideas here are necessarily translations from one community to another. But there are two other, more profound, reasons, one spiritual, the other practical.

The spiritual reason is simple. Personally, I don't believe in magic. That is, there is no series of words, habits or social reference groups that will *make* you do something, just is there isn't any ritual that will make God answer your prayer or deepen your spiritual growth. Instead, I believe in grace and in *mysticism*, an unfortunately similar word to magic, but one with a very different meaning. Mysticism teaches that *God isn't really under human control.* Rather, God decides, on his terms and timing, when to speak and when to help us transform our lives. Often, we simply aren't listening to Him; we miss out and ignore His teachings, big and small. And so, we need to learn to listen. To clear our minds of distraction. To turn our thoughts to God. Even then, there's no magic formula: we can learn from the experience of others, but the journey is our own.

The second reason is quite different. Behavioral scientists know that our best efforts often backfire. The smartest researchers, the most experienced practitioners, still find that their efforts go awry. For example:

- Generally, we can help people take action by automating something they want to do, like setting up automatic deductions into a retirement plan. That is a good thing — except when it backfires and drives people into debt.[115]
- Generally, when people aren't sure what to do, and they see that others are doing something effectively (like saving for the future), a comparison with peers can be quite motivating — except when it backfires and leads people to become discouraged, give up, and save less.[116]
- It can be great to set a diet and refrain from foods we shouldn't eat — until the fact that we've done well with our diet makes us feel that we've "earned" the opportunity to overindulge, and we undo all of the progress we've made. (This is known as *moral licensing*.)[117]

---

[115] Beshears et al. (2017)

[116] Beshears et al. (2015)

[117] E.g., Rubin (2014, 2015)

- Setting clear goals for what we want to accomplish can be very motivating and keep us on track. However, when we've made a mistake or failed to reach a particular milestone, we can throw up our hands and give up trying, or worse, we can backslide because we've failed already (also known as the *What-the-Hell* effect).[118]

What these studies tell us is simply that people and situations differ. What works for one group of people may not work for others. Personal characteristics matter, as do timing and context. What works for one person at one point in time may not help the same person in a different place or time.

And so, for both reasons, we have to learn for ourselves. We have to see what helps us and what doesn't. And we need to resist the temptation to think that *we* have power over spiritual matters. We can't control God, or the Divine, or Allah. We should listen to Him and try our best.

Spiritual design offers ways to try a little harder, perhaps with a bit more success at centering our minds and following our discipline. But there's no guarantee. Many great religious figures — Mother Teresa; John of the Cross — speak of long periods of "dryness" or the "spiritual desert." That's part of accepting that, in the end, we're not in control.

And thus, I hope that ideas here may resonate with you, and help you on your path. But they may not.

Having Faith

Another area I've avoided talking about is how to have more faith; I've focused instead on how to live out your faith. In other words, we've talked about how to turn your intentions into action to pray regularly, to show compassion to your neighbors, or to meditate more deeply on God's word. But these are only relevant if you have faith and the

---

[118] See, for example, Tesser (1996)

intention to act accordingly. And that's not something we can force (and shouldn't try).

For many years in my own life, I struggled with diverging impulses: a sincere lack of belief, and a desire to explore what might be out there. We can't force the former — but we can make it possible for ourselves to do the latter.

Turning our minds to God

Throughout the Bible, both in the New Testament and in the Old Testament, the text enjoins us to "turn to God." The words *turn to* are used as a synonym for *believe in* (or at least that's how modern scholars have interpreted the text). There is certainly an aspect of turning to God which entails turning to His commandments and, as Christians, following Jesus's path. But how do we come to listen to God in the first place, or open ourselves to transformation?

If we have a sincere desire to learn and explore faith, I think there are steps we can take to do so. Certainly, there are examples in the Bible where God has forced someone to pay attention and to listen: Paul's conversion is the most obvious example. But in many others, the progression is different. Consider apostles like Mark: as he deepened his belief and understanding of Jesus, his spirituality deepened over time through new revelation and experiences. A starting point, I think, for intentionally exploring or deepening our faith is that *we open our hearts and be willing to receive*. In other words, we travel an opposite path to that of hardening our hearts.

I don't know of any behavioral research specifically on the opening process and the ability to listen spiritually. There is certainly a tremendous amount of research on the gap between our desire to take action and actually taking action, which we talked about earlier. That section is structured around specific techniques we can use. Here, that's difficult; there's nothing as straightforward as a manual. Instead, I thought I'd try a different route: by describing my own path and my understanding, as a behavioral researcher, and the mechanisms that were at work in opening my heart.

As I look back on my path, I wish that I'd been able to be open to faith and spiritual life sooner. Yet perhaps my winding path, and what I saw along the way, will help you get to where you want to be a bit earlier.

Barriers to openness

In my own path, I think the greatest barrier I've had to faith is that I didn't feel I had any reason to believe. And moreover, I simply didn't think about it.

I have had, and thankfully continue to have, a relatively happy life. I have the good fortune of being born into a life and a time in history that made sense, and provided avenues for advancement, for comfort, and for joy. There was (and is) a ready-made and coherent explanation of the world around me, through a mix of economics, physics, psychology and other sciences. And so you could say I had neither cause to think of faith, nor reason to need it.

Except, of course, for the hunger. The intermittent but persistent hunger. The sense that I had no spiritual center, and that I was missing something I'd never had in the first place.

Because that spiritual hunger came so infrequently, and could be brushed aside or pushed aside, I often did so. But before I did, there were also… moments of openness. I didn't create them, and I certainly didn't take advantage of them.

So how did I become open to that "still, small voice within," as Quakers say? Three factors helped. The first was *mere repetition*. I learned over time that pushing aside that hungry feeling didn't seem to make anything any better. I don't mean that I was unhappy — because I wasn't. Yet even when I was happy, the hunger didn't go away.

Second, I tried to *prepare for those moments of openness, however and whenever they might come*. What do we do when we feel that urge, that hunger, to understand the spirit? For those of us who didn't grow up with a strong education in spiritual practice and organized religion (I didn't start attending any church until high school and didn't receive any real "training" through my unprogrammed Quaker meetings), that's not a trivial question. In my case, I started to slowly educate myself and prepare tools for when these times of interest and openness came.

For example, I got a copy of the Bible, not because I wanted to read it, but because I knew I might want to read it later. I read about how people prayed, not because I wanted to pray, but because I wanted to know how people did it. And I moved frequently when I was a younger man; when I came to a new town, I would look up Quaker meetings in that area. Sometimes I'd go, sometimes I wouldn't. It became a habit of preparation for when I was open to the spirit.

Finally, I tried to *become familiar with the routines of faith*. That included going to Meeting and, later, sitting in silence before each meal. I would try these things, not because I directly felt the Spirit, but because I didn't want foreignness or unfamiliarity to be a barrier.

For many of us who don't grow up in a strong faith community, there may be many reasons we're closed off from faith and spiritual practice. One is a sincere and honest belief that religion and spirituality aren't demonstrated, proven, and real. That's not something we should seek to overcome or force; faith is faith. But there are other factors that block us from exploring, from answering that hunger within (if and when it comes).

One factor is our social norms: Discussion of spirituality has become so uncommon in our culture that it seems unusual and foreign to pursue it ourselves.[119] Our fear of seeming unusual or different from our friends isn't rooted in whether God exists or no; it's simply a social fear.

Another factor is that we don't know how to answer that hunger, and it feels uncomfortable and difficult to try. When those moments come, we don't have the tools to respond. Taking the time to pray or meditate *feels odd* in part because we've never done it before: it doesn't come to mind easily (the availability heuristic), isn't something encountered often (the mere exposure effect), and we feel uncomfortable because

---

[119] Merritt (2017)

we don't know how to do it well (a lack of self-efficacy). These are reasons that shouldn't matter — but do.

As a behavioral scientist, I study such barriers as these in a non-spiritual context, in an attempt to figure out how to overcome them. So, personally, I tried to separate my *lack of faith* when I was younger from my *unfamiliarity with faith* and its routines. I sought to address the latter, to give the former a fairer chance of being assessed and explored on its own terms.

Each person, of course, will have their own techniques or actions that make them feel a little bit more open and willing to listen. For some, that might be taking walks in nature. For others it might be meditation. It might even be looking out over the vastness of a crowd.

When discover or want to try acts that can help us listen and explore our faith, we can use the techniques discussed throughout this book in the context of spiritual practice. In other words, the same techniques that can help us act according to our faith can help us do the things that aid us in unhardening our hearts. We can't force or use behavioral design to instill faith itself, but we can design our environment to help us listen for faith's call, if and when it comes.

Building awareness into each moment

Part of being open to faith's call is to build awareness of spiritual matters into our daily lives: of the beautiful creation around us, of God Himself, of our hunger for spiritual growth.

How can we bring about deliberative awareness in the moment? One way is to *build contemplative habits into our daily routine*. The habit itself is something we automatically do, like praying before meal, or saying a short prayer when we pass a church. Like any habit, it depends on a consistent and regular trigger: seeing the plate of food in front of oneself, or driving by the church. But unlike normal habits, in this one the routine we invoke upon seeing that trigger is quite different, *because it is meant to clear our minds rather than do a specific thing*. The idea is to bring our spiritual identity to the fore — to create a purposeful pause in our day, and to spend time on spiritual matters.

Spiritual Design

Another technique to encourage deliberate thought (and hopefully contemplative thought), is to *intentionally add friction throughout one's life*. We slow ourselves down so that we cannot take rapid and habitual action. What would that look like in general? Moving very slowly. Pausing before answering any question, or responding to any remark. Taking a small mental break before doing anything that feels important.

Mindfulness, or a constant awareness of the present moment, is central to Buddhist practice. It can be of inspiration whether one is Buddhist or not. For example, Buddhist practice can include focusing on one's breathing, using verbal and mental tools like mantras to clear the mind, and watching our thoughts flow by without disruption or engagement. Each of these can help us focus on the moment, and our path toward God.

Faith, Design, and our Children

If it is possible to prepare the way for ourselves to explore and be open to receiving God's message, what does that mean for how we raise our children? Once we try to apply these techniques to another person, a key element of the process — our yearning to reach the Divine — is potentially lost.

Spiritual design is based on a sincere and willing feedback loop: a person changes their environment, sees if that brings them closer to the Divine, and learns from that experience, adapting and making changes over time. It's an iterative process in which the person repeatedly reaches out to the Divine to listen, to commune, and to learn His will. There's no feedback loop if the person doesn't *want* to do this. And if that inner yearning isn't already there, we risk coercing our children into an empty ritual — something that makes it harder for them to find God.

Some aspects of spiritual design may be reasonable and helpful for our children, however. One of the most basic techniques is simply education: knowing about God and his teachings, as a guide to help people seek Him out when they so choose. Many faiths believe in the ever-present opportunity of connecting with the Divine; it's the daily distractions of life that make it hard for us to find the way on our own. A little education about what's possible can open the door for our children, without pushing them through it before they are ready.

In this way, basic knowledge of scripture and of spiritual practice can be a spiritual tool. It's something children can learn and deploy in their lives when and how they need. Regardless of where they are in their spiritual path, and their relationship with God, knowing that the scripture and other teachings are there can help them explore further. Similarly, knowing about the Church, and the community of believers who support and welcome people who seek closer communion with God, can be a tool for children to use in the future. They may choose never to use that tool, or it may help them when they need it. If we teach our children about prayer, and how it can work and enrich our lives, we also arm them with a spiritual tool. We leave room for them to employ that tool in their own authentic search and connection with God.

Trouble can arise when we try to *instill* spiritual habits. As we discussed earlier, habits are funny things: they are nonconscious and free our minds to do other things. The more we repeat something, the more our minds say, *Okay this is something I need to do often; let me make it effortless and mindless.* Let's think about eating. We've all had the experience of only barely "tasting" or "experiencing" the food we're eating because our minds are focused on something completely different while we shovel it into our mouths. The more we eat the same thing in the same way, the less special and noticeable it is.

If we don't give particular attention to God during them, our spiritual rituals can become like mindless eating: just something we do because we have to, while we're thinking about something else. That isn't necessary of course; habits (and the spiritual rituals they support) help us clear and focus our minds on God. But that works *if we're searching and yearning for God.* Yearning for God isn't something we can force upon our children. They themselves determine if prayer and other spiritual rituals are meaningful. We can only supply them with the tools and the opportunity to use them.

Beyond habits and spiritual tools, as parents we also play a major and beneficial role in setting the norms of behavior. Especially before their children's teenage years, parents have a strong say in what other children their kids play with and hang out with. If the acts of finding faith are like other behaviors — health behaviors, eating behaviors and studying behaviors — then social group matters. Kids who are around others who study tend to study more. Kids who are around others who exercise tend to exercise more. And if spiritual behavior is similar — a big *if*, but one to consider — kids who are around others who pray (outward behavior) and talk about their search for God (inward behavior) may be more likely to pray and search for God themselves.

Personally, my wife and I considered this social effect of spirituality very strongly when we thought about where to put our kids in school. The Quaker Meeting my wife and I were attending when our children were born had a school associated with it: Sandy Spring Friends School. Academically, it's a very good school, from what we could see. More importantly though, some of the children from the school also attended our church. And they were good kids: confident, well behaved, thoughtful, and regularly attending Sunday school and participating in church. We wanted our kids to be like them and explored ways to send our kids to school there; unfortunately, we moved out of the area before it was financially possible for us. After we moved to Chicago, a major factor in joining a church was its children's community, so our kids could have a supportive environment.

Sending kids to a religious school is one way to shape the behavioral norms a child encounters. I know some people reading this are reacting in alarm, if not outright anger — especially people who were raised in religious schools and felt coerced or abused there. Their experiences should be a warning sign to all well-meaning parents: a religious school doesn't magically mean a good environment for kids. Kids learn from other kids, and from their teacher. If those two groups are setting a bad example, kids will learn that, regardless of the lofty principles on which the school is founded. Further, if the teachers enforce habits of the spirit without supporting children to find their own authentic connection with God, they can do real spiritual harm. Children may rightfully see their experience of religious expression as empty, and turn away from God. On their own, I would argue that religious schools can help, and are a non-coercive action for parents to take — but parents can't turn a blind eye to the coercion and examples being set by the members of the school community itself.

And, finally, parents help shape the identity of a child, and that identity can have a spiritual character that makes it more natural for the child (as children or as adults) to find faith themselves — or not. For example: parents who instill a strong sense of atheism into their kids, and define them as "atheist," just as another parent would label and inspire in their kid an identity as a "German-American" or "Hispanic-American." Identity, as we talked about earlier, is a powerful force that guides one's behavior in new or ambiguous situations. If I'm visiting a new city and walk down a street where I encounter a beautiful old church, whether or not I go in depends, in part, on whether I see myself as a religious person (and a person who likes old buildings). If I see myself as an atheist, I just wouldn't pay as much attention to the beautiful old church, and to me it wouldn't feel it natural to walk in, sit and meditate for a while. Identity matters. Of course, as parents of teenagers well know, identity is not something parents can force upon their children. Like prayer, knowledge of the Bible, and the social norms of faith, it is something that parents can *offer* to their children, and pray that they choose to incorporate and build upon in their lives.

Quakers and Children

The dangers of empty ritual are something my Quaker forebears recognized and forcefully fought against. They saw an undue focus on external sacraments and ceremony as distracting from the inner communion with God. They resisted having Sunday School, or any formal teaching of the Bible. They didn't employ a behavioralist's understanding of mindless habits, but their descriptions and concerns are remarkably similar. And so, they removed distractions in the name of one's inner religious life.

Some later Quakers believed this approach went too far: it not only removed the repetition of habit, it also removed the opportunity for Quaker children and newcomers to the faith to learn about the Bible and be armed with spiritual tools they could use to build communion with God. The vast majority of Quaker meetings now have both Sunday School, a sermon, and a teaching of or reference to the Bible, in part for these reasons. The Quaker example is one that I think is particularly helpful to delve into further to understand the power, and the limits, of spiritual design.

Reflections on Quaker Spiritual Design

Early on in this book, I wrote about how the Quakers designed their space, and their worship, purposefully. To better understand why and how that happened, a little history and background might help.

Brief history of Quakerism

The Quakers, or more formally, the Religious Society of Friends, started in the mid-1600s in England, during a period of tremendous upheaval. England fought a series of civil wars, and numerous religious movements challenged the supremacy of the Episcopal Church. A band of likeminded people came together under the banner of the Religious Society of Friends in Northeast England during this time. While the traveling preacher George Fox was one of the best known of the group, they did not have a single "leader" in any formal or theological sense. The Society rejected the outward rituals and "notions" of the Protestant Church at the time and focused instead on each person's ability to directly connect with God and be guided by the Holy Spirit.

Since the mid-1600s, the Religious Society of Friends has spread all over the world, with the largest concentrations in East Africa, the United States, and South America. While there is a great diversity within, Quakers are commonly known for a few characteristics:

- Quakers do not have a "creed" per se. There is no single accepted statement of Quaker beliefs.
- The "bureaucracy" of the church is very limited, if it could be said to exist at all; many unprogrammed meetings have no formal preachers or spiritual teachers, for example.
- Instead of a written creed or church authority, Quakers believe that we each have the ability to listen to the Divine — often referred to as the Inward Light, inner Teacher, or Christ Within.
- Quakers believe that our actions in daily life matter. Quakers should "let their lives speak," following the example set by Jesus Christ and his Apostles. A true Christian is one who not only professes faith in God, but also puts his teachings into practice in daily life.
- Because of this attention to putting God's teachings into practice, Quakers have developed a set of *testimonies*, or common understandings of what it means to follow the Inward Light. Currently these include:
    a. Simplicity — living with restraint and moderation in dress, in action, and in speech.
    b. Peace — renouncing war and seeking an end to conflict among people.
    c. Integrity — consistency in word and deed.
    d. Community — turning to one's religious community to discern the leadings of God.
    e. Equality — treating others as equal, regardless of outward characteristics or their standing in society.

In addition to these "additive" elements of Quakerism, early Quakers rejected what they considered distracting elements of Christian practice at the time. For example, early Quakers:

- Built Meeting Houses (churches) that were largely unadorned of outward symbols of religion;

- Radically changed the experience of church: people prayed in silence, waiting for the Lord to speak to them, without a structured service, preacher or ceremony;
- Resisted having Sunday School or other forms of religious teaching; leading Quakers knew the Bible by heart, so "teaching" the text was largely unnecessary;
- Did not practice water baptism, communion or other outward sacraments;
- Did not celebrate religious festivals such as Christmas and Easter (since Christ's birth and death should be celebrated every day) and renamed the days of the week to a "plain calendar": first/second/third/etc. day, first/second/third/etc. month.

The result was outward austerity, in the name of inner richness of religious life. Nothing should stand in the way of direct communion with God, nor distract the believer with earthly matters.

Quaker testimonies and outward austerity also set Quakers apart as a community: Quakers adopted distinctive "plain" dress that visually made them different than other members of society. In the name of equality before God, they refused to use special language and rituals, like taking off one's hat to honor upper class members of society. They refused to swear oaths (as Jesus indicated in Matthew 5:34), or fight in wars (Matthew 5:9). Quakers looked, spoke, acted and worshiped differently than their peers.

Not all of these unique characteristics, or "distinctives" as they are known, survived over the last 350 years, but many have. In the unprogrammed Quaker tradition that I come from, worship is similar to the early Quaker practice. Individuals enter the room at the appointed time and wait in silence for the Lord to guide people to speak. A person who is led to do so stands, speaks, then sits down into the silence. Otherwise, there is no programming or structure. Even in programmed worship though (which incorporates a sermon and potentially other normative elements of a church service, such as announcements, singing, vocal prayer, etc.), many services maintain a focus on "open worship," or sitting in silence for the Lord.

The testimonies have evolved over time, but have largely remained. Meeting houses tend to be unadorned, simple buildings. Outward sacraments are still rare across Quakerism. Sunday school is more common though, especially among programmed meetings. Quakers do, by and large, celebrate Christmas and Easter, and largely have abandoned a distinct dress and style of speech.

Since unprogrammed meetings are *outwardly* distinctive, and are the tradition I know best, let's analyze what that distinctiveness does.

How Quaker distinctives shaped the faith

For early Quakers, the removal of the outward rituals of faith allowed them to focus on their rich inner religious lives. As noted above, many leading Quakers knew the Bible by heart, and thus did not need a preacher to "teach" them the Bible. Since Quakers believe in direct revelation, they did not see the need for a preacher to interpret the Bible either. In their inward communion with God, Quakers felt they were overcome with the power of the Holy Spirit, guided to speak, and they literally quaked with religious passion (hence the name *Quakers*). They were steeped in Christianity, and felt a deep connection with God — thus it was quite reasonable to see outward rituals and forms as a distraction.

They constructed their outward environments to remind themselves of their inward communion: intentional plainness, intentional lack of structure or ritual, intentional simplicity. They structured their "non-ritual rituals" likewise to express, and reinforce, their testimonies, refusing to use honorific titles, or giving hat-honor.

Because of their outward distinctiveness, they also created a clear and separate community of believers. Quakers could recognize each other in daily life, and thus see their belief in action, despite being part of a much-larger, non-Quaker population. In other words, they created a strong normative community.

Unintended consequences

In today's world, the context of spiritual practice is quite different, especially for Quakers in the unprogrammed tradition. In my personal experience, the very characteristics that made Quakers distinctive in the early years can make it difficult to learn Quakerism now.

For example, the lack of a preacher means that it is very difficult for new people to gain a deep understanding of the richness of Quaker tradition and belief. In the unprogrammed meetings that I've attended, a thoughtful rejection of creeds and strict doctrine means that religious education for new members and children is limited, and ecumenical in the extreme, often not teaching the Bible or historical Quaker interpretations of it and focusing instead on a sampling of world religions.

Similarly, Quaker silence is an amazing force and intentional tool to focus on the Divine. As William Penn wrote:

> True silence is the rest of the mind; and is to the spirit what sleep is to the body, nourishment and refreshment.

However, as with many good things, there are non-obvious effects. Since worshipers in an unprogrammed meeting are silent most of the time, it's difficult for new people to know what to expect and what is normative. When people stand to speak, the newcomer (or child) would not know that, according to historical Quakers, the message should only be spoken if it is guided by God. The outward form simply looks like someone wanting to say something, saying it, and then moving on.

In many ways, the intentional choices of design that *supported* deep spiritual worship for early Quakers, and modern Quakers already steeped in the tradition, may unfortunately *limit* or *hinder* the passing on of that tradition to new people. In order words, the best of our intentions can have unintended consequences — not ones that are fixed in stone and immutable, but ones that we can recognize and work to mitigate. The Quaker story is one of many one could tell about how communal (and individual) worship in religious communities can have both great benefits and less-welcome unintended consequences. It's simply the one that I know well, and one that can illustrate the power and limitations of spiritual design.

As we find in the broader behavioral literature, even the most wonderful aspects of an intentionally designed environment can affect people in different ways, and can even affect the same person differently over time. In the design of communal worship, this reminds us of the need for a certain type of vigilance: to watch over the spiritual lives of members, and to see how the communities' traditions interact with and serve the needs of its people, and how a range of programs might be weaved together to address the diverse needs of the community. Or, more simply: we shouldn't expect one size to fit all.

At an individual level, the lesson is somewhat different. As we seek to support spiritual growth and transformation in ourselves, we should expect that some techniques and traditions will serve us well, and others won't. There is nothing wrong with either us or the traditions we draw from if that is the case. A particular facet of our tradition may not serve us well in the moment, but in another context and time it may provide the life-giving sustenance we need. We should expect a diversity of experience, and we should always keep our minds focused on the purpose of spiritual design: enriching our spiritual growth and practice. That involves experimentation, openness and a willingness to accept that any particular approach may not aid us on our path — at least, not right now.

10
CLOSING THOUGHTS

May these words of my mouth and this meditation of my heart be pleasing in your sight, LORD. — Psalm 19:14

This book is both a description of spiritual design and a personal expression of it. Two and half years ago, when I started writing this, I did so because I wanted to delve deeper into my faith. I wanted to learn more about God, about prayer, and about how my environment has shaped, and could shape, my ability to be a more faithful spiritual person. I learn through writing.

I also knew that through writing, and talking about these topics with others, I might keep myself from 'backsliding': from returning to the easy comfort of a modern life that's devoid of self-reflection and hard spiritual questions. My social commitment around the book, my self-signaling, my gaining comfort with a new spiritual language, and my establishment of regular spiritual contemplation, all of which are part of the writing of this book, were, well, intentional spiritual design in themselves.

And it went as all human endeavors do: imperfectly. I was afraid of talking about this book and sharing it with others, fearing how others might judge my spiritual path and my naivety. And I was lazy: I set a clear schedule to keep on track and reinforce my spiritual discipline, and I failed to keep it up. In countless moments, I had the choice to fixate my eyes on God through writing and prayer; I chose to watch Amazon Prime TV instead.

But yet it's helped. My friends asked me about the book, and what happened to it. The stack of books I bought for background research stood as a physical reminder of what I wanted to accomplish, and what it meant to me. And over time, the hunger for meaning and God's truth, along with some annoying calendar reminders and half-built habits, overcame the "play" button on Amazon Prime Video (on some nights, at least).

This isn't a story of personal triumph. Instead, it's how I've used spiritual design in my own life is a story of imperfect, halting attempts to get closer to God. It's a small part of "doing my part." I couldn't compel spiritual growth any more than anyone else can. But in writing, and in applying spiritual design in my own life, I've tried to show up, and be a bit more open and willing to hear God's voice.

Such imperfection is what we should expect. We're human. We should also expect that, on the margin, we can make things a bit better if we try, if we persist and have faith, and if we listen.

A Review of the Major Lessons

As we look back across the range of research studies and techniques discussed, we've covered quite a bit of ground, from fighting distraction and procrastination to overcoming temptations and bad habits. To reinforce these lessons, here is a quick summary of the underlying themes:

- We are profoundly affected by our environment: our physical environment of sights and sounds, our internal environment of hunger and yearning, and our social environment of friends and expectations.
- Sometimes that environment supports us and sometimes it hinders us from doing the things we want to do, and from doing the things we know we should do.
- We can change our environment. We can intentionally design it to support our spiritual path: that's Spiritual Design. We can change our environment, both to encourage us to spend time in meditation and with God and God's word, and to help us more fully let our lives speak through faithful action and acts of compassion.
- Spiritual design will be an uphill battle. Our environment isn't a neutral thing: Others are very intentionally designing our environment to pull our attention to their ads, their products and their solutions to our secular problems. That's how they make money.
- But in the end, we have a tremendous set of tools to:
    1) Make it **E**asier to live out our spiritual lives, by using defaults and removal of friction, celebrations of small wins, and planning through the obstacles that stand in our way.
    2) Grab **A**ttention to what matters in our faith and practice, with powerful cues, simple reminders, a wise understanding of when we have attention to spare and mental bandwidth, and when we can build on existing habits and routines to get us started.
    3) Embrace the **S**ocial nature of human beings, by intentionally establishing a spiritually inspiring reference group, seeking out a community of others who will support us, and intentionally building social commitments to follow through on our path.

> 4) Find the urgency to act in a **T**imely manner, by attaching what we want to do to an already urgent action, by using prospective hindsight to experience our future regret and joy, and by creating mental milestones and deadlines to mark our progress along the way.

- As we progress on our spiritual path, we can work to lock in our changes through intentionally building habits, recasting our self-narrative, setting up clear feedback loops, and forgiving ourselves when we falter.
- Sometimes, though, we have to *stop* in order to *start*. We have to stop habits and behaviors that lead us down the wrong path. Here again, we can find hints of an answer, in looking to the ways habits are triggered, avoiding, hijacking and redirecting those triggers, or embracing mindfulness in our daily lives.
- We can also engineer spiritual design in reverse: purposefully making it hard to stray from the better path, once chosen.

Spiritual design isn't a panacea. It won't *make* you do anything. But it could help you do what you already want to do: to reinforce your good tendencies, and help let go of some of the bad ones. The goal is to help you, in a bit more of your life, and in a bit more of your heart, become closer to God.

An Open Call for Research

As I've tried to make clear throughout this book, there is little research on the intersection of behavioral science and spirituality. This book seeks to bring together those two communities by interpreting the lessons from one in the context of the other. But that isn't the same as research directly focused on the topic.

There certainly is scientific research about religion and spirituality, but much of it is on whether science can "explain" religion or whether religion can withstand the onslaught of scientific discoveries. This research can make it feel like spirituality is on the defensive against scientific inquiry. Certainly, many people in the scientific study of religion are, in fact, explicitly seeking to undermine religious beliefs and experiences.

It needn't be that way, however. Thoughtful scientists and researchers throughout the centuries have found that scientific inquiry can help them reveal the wonder of God's creation. And, more broadly, that scientific inquiry can live side-by-side with a deep sense of spirituality and belief.

In that vein, *I'd like to make an open call for serious and focused research on the behavioral science of spirituality*. Behavioral science has a particular understanding of the limitations of human beings — one that aligns quite well with the depictions of humanity and our struggles that are found in the Bible and in other religious texts. And, like many religious traditions, many behavioral scientists believe that with forethought and good will people can do a bit better. People can make better decisions, be clearer in their thinking, and follow through on their commitments to act.

This new research into behavioral science and spirituality would help us understand:

- *What factors in our environment best help us follow through on our spiritual commitments?*
- *What role do spiritual communities play in supporting, or undermining, our spiritual growth?*
- *How do different people respond to the diverse forms of prayer, as a means for them to seek out and connect with the Divine?*

Personally, I believe that the final answers to these questions are in God's hands. But that that doesn't mean we shouldn't try to explore them and seek to better understand ourselves and our behavior. As spiritual people, we have a responsibility to turn our minds to matters of faith as best we can: to strive to follow to our religious teachings, even when we know that we will fall short of our ideals. As I mentioned in the Introduction, the fact that God's grace is necessary does not absolve us of the need to do our part.

This book, again, is a translation of behavioral science into a spiritual domain. New research could allow us to write this story without translation, and in the native language of spirituality: through field research and experiments, with willing participants who want to deepen their spiritual lives. This work could help us understand what seems to create additional roadblocks to our spiritual paths, and what seems to remove them.

Because spiritual questions have been considered "off-limits" for serious inquiry, a vital part of people's lives hasn't gained from thoughtful and sincere attention. That, I believe, should change. We can treat our spiritual lives — especially the intention-action gap between what we want to do and what we actually do — as we do any other part of our lives: something to be learned about, reflected upon, and, with grace, made slightly better over time.

BIBLIOGRAPHY

Alba, Joseph W. *Consumer Insights: Findings from Behavioral Research*. Edited by Joseph W. Alba. First. Cambridge, Mass: Marketing Science Institute, 2011.
Andreoni, James, Justin M. Rao, and Hannah Trachtman. "Avoiding the Ask: A Field Experiment on Altruism, Empathy, and Charitable Giving." *Journal of Political Economy* 125, no. 3 (April 28, 2017): 625–53.
Appiah, Anthony. *Experiments in Ethics*. Harvard University Press, 2008.
Ariely, Dan. *Predictably Irrational: The Hidden Forces That Shape Our Decisions*. New York, NY: HarperCollins, 2008.
———. *The Honest Truth About Dishonesty: How We Lie to Everyone—Especially Ourselves*. New York, NY: Harper Perennial, 2013.
Ashraf, Nava, Dean Karlan, and Wesley Yin. "Tying Odysseus to the Mast: Evidence from a Commitment Savings Product in the Philippines." *The Quarterly Journal of Economics* 121, no. 2 (2006): 635–72.
Augustine. *Confessions*. Translated by Sarah Ruden. Translation edition. New York, NY: Modern Library, 2018.
Baer, Marc David, Todd Michael Johnson, Lily Kong, Seeta Nair, Henri Paul Pierre Gooren, Peter G. Stromberg, and Fenggang Yang. *The Oxford Handbook of Religious Conversion*. Oxford University Press, 2014.
Baer, Ruth A. "Mindfulness Training as a Clinical Intervention: A Conceptual and Empirical Review." *Clinical Psychology: Science and Practice* 10, no. 2 (2003): 125–143.
Bamberg, Sebastian. "Is a Residential Relocation a Good Opportunity to Change People's Travel Behavior? Results From a Theory-Driven Intervention Study." *Environment and Behavior* 38, no. 6 (November 1, 2006): 820–40.
Behavioural Insights Team. "EAST: Four Simple Ways to Apply Behavioural Insights," 2014. http://www.behaviouralinsights.co.uk/publications/east-four-simple-ways-to-apply-behavioural-insights/.
———. "Behavioural Insights in Australia," November 26, 2017. https://www.bi.team/blogs/behavioural-insights-in-australia/.

Bekkers, René, and Pamala Wiepking. "A Literature Review of Empirical Studies of Philanthropy: Eight Mechanisms That Drive Charitable Giving." *Nonprofit and Voluntary Sector Quarterly* 40, no. 5 (October 1, 2011): 924–73.

Benartzi, Shlomo. *Thinking Smarter: Seven Steps to Your Fulfilling Retirement…and Life*. New York, NY: Portfolio, 2015.

Berns, Gregory S., Jonathan Chappelow, Caroline F. Zink, Giuseppe Pagnoni, Megan E. Martin-Skurski, and Jim Richards. "Neurobiological Correlates of Social Conformity and Independence During Mental Rotation." *Biological Psychiatry* 58, no. 3 (August 1, 2005): 245–53.

Beshears, John, James J. Choi, David Laibson, Brigitte C. Madrian, and Katherine L. Milkman. "The Effect of Providing Peer Information on Retirement Savings Decisions." *The Journal of Finance* 70, no. 3 (2015): 1161–1201.

Beshears, John, James J. Choi, David Laibson, Brigitte C. Madrian, and William L Skimmyhorn. "Borrowing to Save? The Impact of Automatic Enrollment on Debt." Working Paper, 2017. https://scholar.harvard.edu/laibson/publications/borrowing-save-impact-automatic-enrollment-debt.

Bettinger, Eric P., Bridget Terry Long, Philip Oreopoulos, and Lisa Sanbonmatsu. "The Role of Application Assistance and Information in College Decisions: Results from the H&R Block FAFSA Experiment." *The Quarterly Journal of Economics* 127, no. 3 (August 1, 2012): 1205–42.

Bodner, Ronit, and Drazen Prelec. "Self-Signaling and Diagnostic Utility in Everyday Decision Making." *The Psychology of Economic Decisions* 1 (2003): 105–26.

Castillo, Marco, Ragan Petrie, and Clarence Wardell. "Fundraising through Online Social Networks: A Field Experiment on Peer-to-Peer Solicitation." *Journal of Public Economics* 114 (June 1, 2014): 29–35.

Chatzisarantis, Nikos L. D., and Martin S. Hagger. "Mindfulness and the Intention-Behavior Relationship Within the Theory of Planned Behavior." *Personality and Social Psychology Bulletin* 33, no. 5 (May 1, 2007): 663–76.

Choi, James J, David Laibson, Brigitte C Madrian, and Andrew Metrick. "Defined Contribution Pensions: Plan Rules, Participant Choices, and the Path of Least Resistance." In *Tax Policy and the Economy*, 16:67–114. MIT Press, 2002.

Clear, James. *Atomic Habits*. New York, NY: Random House Business Books, 2018.

Covey, R. Stephen. *The 7 Habits of Highly Effective People*. New York: Simon & Schuster, 2013.

Dai, Hengchen, Katherine L. Milkman, and Jason Riis. "The Fresh Start Effect: Temporal Landmarks Motivate Aspirational Behavior." *Management Science* 60, no. 10 (June 23, 2014): 2563–82.

Damasio, AR, BJ Everitt, and D Bishop. "The Somatic Marker Hypothesis and the Possible Functions of the Prefrontal Cortex [and Discussion]." *Philosophical Transactions: Biological Sciences* 351, no. 1346 (1996): 1413–20.

Darley, John M, and C Daniel Batson. "'From Jerusalem to Jericho': A Study of Situational and Dispositional Variables in Helping Behavior." *Journal of Personality and Social Psychology* 27, no. 1 (1973): 100.

Dean, Jeremy. *Making Habits, Breaking Habits: Why We Do Things, Why We Don't, and How to Make Any Change Stick*. Boston, MA, USA: Da Capo Press, 2013.

Duckworth, Angela L., Tamar Szabó Gendler, and James J. Gross. "Situational Strategies for Self-Control." *Perspectives on Psychological Science* 11, no. 1 (January 1, 2016): 35–55.

Duhigg, Charles. *The Power of Habit: Why We Do What We Do in Life and Business*. New York, NY: Random House, 2012.

Engber, Daniel. "Everything Is Crumbling." *Slate*, March 6, 2016. http://www.slate.com/articles/health_and_science/cover_story/2016/03/ego_depletion_an_influential_theory_in_psychology_may_have_just_been_debunked.html.

Evers-Hood, Ken. *The Irrational Jesus: Leading the Fully Human Church*. Eugene, Oregon: Cascade Books, 2016.

Eyal, Nir. *Indistractable: How to Control Your Attention and Choose Your Life*. Dallas, TX: BenBella Books, 2019.

Fertig, Andrew, Jaclyn Lefkowitz, and Alissa Fishbane. "Using Behavioral Science to Increase Retirement Savings." *Ideas42 Whitepaper*, 2018. https://www.ideas42.org/wp-content/uploads/2018/11/I42-1046_MetLifeLatAm_paper_ENG_Final.pdf.

Fogg, BJ. "Tiny Habits." Accessed May 27, 2019. https://www.tinyhabits.com.

———. *Tiny Habits: The Small Changes That Change Everything*. Boston: Houghton Mifflin Harcourt, forthcoming.

Galla, Brian M., and Angela L. Duckworth. "More than Resisting Temptation: Beneficial Habits Mediate the Relationship between Self-Control and Positive Life Outcomes." *Journal of Personality and Social Psychology* 109, no. 3 (September 2015): 508–25.

Gardner, Ben D. "Busting the 21 Days Habit Formation Myth | UCL 'Health Chatter': Research Department of Behavioural Science and Health Blog," 2012. http://blogs.ucl.ac.uk/bsh/2012/06/29/busting-the-21-days-habit-formation-myth/.

Gerber, Alan S., and Donald P. Green. *Field Experiments: Design, Analysis, and Interpretation*. New York, NY: W.W. Norton, 2012.

Gerber, Alan S., and Todd Rogers. "Descriptive Social Norms and Motivation to Vote: Everybody's Voting and so Should You." *The Journal of Politics* 71, no. 01 (2009): 178–91.

Goldstein, Daniel G., Hal E. Hershfield, and Shlomo Benartzi. "The Illusion of Wealth and Its Reversal." *Journal of Marketing Research* 53, no. 5 (December 11, 2015): 804–13.

Hamilton, Jon. "Think You're Multitasking? Think Again." NPR.org, 2008. http://www.npr.org/templates/story/story.php?storyId=95256794.

Hayes, Steven C., Kirk D. Strosahl, and Kelly G. Wilson. *Acceptance and Commitment Therapy: The Process and Practice of Mindful Change*. New York: Guilford Press, 2011.

Hershfield, Hal E, Daniel G Goldstein, William F Sharpe, Jesse Fox, Leo Yeykelis, Laura L Carstensen, and Jeremy N Bailenson. "Increasing Saving Behavior Through Age-Progressed Renderings of the Future Self." *Journal of Marketing Research* 48, no. SPL (November 2011): S23–37.

Hershfield, Hal E., Elicia M. John, and Joseph S. Reiff. "Using Vividness Interventions to Improve Financial Decision Making." *Policy Insights from the Behavioral and Brain Sciences* 5, no. 2 (October 2018): 209–15.

Himle, Michael B., Douglas W. Woods, John C. Piacentini, and John T. Walkup. "Brief Review of Habit Reversal Training for Tourette Syndrome." *Journal of Child Neurology* 21, no. 8 (August 1, 2006): 719–25.

Hofmann, Wilhelm, Roy F. Baumeister, Georg Förster, and Kathleen D. Vohs. "Everyday Temptations: An Experience Sampling Study of Desire, Conflict, and Self-Control." *Journal of Personality and Social Psychology* 102, no. 6 (June 2012): 1318–35.

Imhoff, Roland, Alexander F. Schmidt, and Friederike Gerstenberg. "Exploring the Interplay of Trait Self-Control and Ego Depletion: Empirical Evidence for Ironic Effects." *European Journal of Personality* 28, no. 5 (September 1, 2014): 413–24.

Iyengar, Sheena S. *The Art of Choosing*. New York, NY: Hachette Book Group, 2010.

Johnson, Eric J., and Daniel Goldstein. "Do Defaults Save Lives?" *Science* 302, no. 5649 (November 21, 2003): 1338–39.

Jones, James W. *Can Science Explain Religion?: The Cognitive Science Debate*. Oxford; New York: Oxford University Press, 2015.

Kahneman, Daniel. "Evaluation by Moments: Past and Future." *Choices, Values, and Frames*, 2000, 693–708.

———. *Thinking, Fast and Slow*. New York, NY: Farrar, Straus and Giroux, 2011.

Karlan, Dean, and John A. List. "Does Price Matter in Charitable Giving? Evidence from a Large-Scale Natural Field Experiment." *American Economic Review* 97, no. 5 (December 2007): 1774–93.

Karlan, Dean, Margaret McConnell, Sendhil Mullainathan, and Jonathan Zinman. "Getting to the Top of Mind: How Reminders Increase

Saving." National Bureau of Economic Research Working Paper, 2010.http://www.dartmouth.edu/~jzinman/Papers/Top%20of%20 Mind%202011jan.pdf.

Kavanagh, David J, Jackie Andrade, and Jon May. "Imaginary Relish and Exquisite Torture: The Elaborated Intrusion Theory of Desire." *Psychological Review* 112, no. 2 (2005): 446.

Keller, Punam Anand, Bari Harlam, George Loewenstein, and Kevin G. Volpp. "Enhanced Active Choice: A New Method to Motivate Behavior Change." *Journal of Consumer Psychology*, Special Issue on the Application of Behavioral Decision Theory, 21, no. 4 (October 2011): 376–83.

Khoury, Bassam, Tania Lecomte, Guillaume Fortin, Marjolaine Masse, Phillip Therien, Vanessa Bouchard, Marie-Andrée Chapleau, Karine Paquin, and Stefan G. Hofmann. "Mindfulness-Based Therapy: A Comprehensive Meta-Analysis." *Clinical Psychology Review* 33, no. 6 (August 1, 2013): 763–71.

Kühberger, Anton, and Carmen Tanner. "Risky Choice Framing: Task Versions and a Comparison of Prospect Theory and Fuzzy-Trace Theory." *Journal of Behavioral Decision Making* 23, no. 3 (2010): 314–29.

Lally, Phillippa, Cornelia H. M. van Jaarsveld, Henry W. W. Potts, and Jane Wardle. "How Are Habits Formed: Modelling Habit Formation in the Real World." *European Journal of Social Psychology* 40, no. 6 (2010): 998–1009.

Langer, Ellen J., Arthur Blank, and Benzion Chanowitz. "The Mindlessness of Ostensibly Thoughtful Action: The Role of 'Placebic' Information in Interpersonal Interaction." *Journal of Personality and Social Psychology* 36, no. 6 (1978): 635–42.

Latané, Bibb, and John M. Darley. *The Unresponsive Bystander: Why Doesn't He Help?* New York: Appleton-Century-Crofts, 1970.

Loewenstein, George. "Out of Control: Visceral Influences on Behavior." *Organizational Behavior and Human Decision Processes* 65, no. 3 (March 1, 1996): 272–92.

Manis, Melvin, Jonathan Shedler, John Jonides, and Thomas E. Nelson. "Availability Heuristic in Judgments of Set Size and Frequency of Occurrence." *Journal of Personality and Social Psychology* 65, no. 3 (1993): 448–57.

Merritt, Jonathan. *Learning to Speak God from Scratch: Why Sacred Words Are Vanishing—and How We Can Revive Them.* New York: Convergent Books, 2018.

Merton, Thomas, and Sue Monk Kidd. *New Seeds of Contemplation.* New York: New Directions, 2007.

Milkman, Katherine L., Julia A. Minson, and Kevin Volpp. "Holding the Hunger Games Hostage at the Gym: An Evaluation of Temptation Bundling." SSRN Scholarly Paper. Rochester, NY: Social Science Research Network, April 3, 2013. http://papers.ssrn.com/abstract=2183859.

Miller, George. "The Magical Number Seven, plus or Minus Two: Some Limits on Our Capacity for Processing Information." *The Psychological Review* 63 (1956): 81–97.

Mischel, Walter, Ebbe B. Ebbesen, and Antonette Raskoff Zeiss. "Cognitive and Attentional Mechanisms in Delay of Gratification." *Journal of Personality and Social Psychology* 21, no. 2 (1972): 204–18.

Mitchell, Deborah J., J. Edward Russo, and Nancy Pennington. "Back to the Future: Temporal Perspective in the Explanation of Events." *Journal of Behavioral Decision Making* 2, no. 1 (January 1, 1989): 25–38.

Nickerson, Raymond S. "Confirmation Bias: A Ubiquitous Phenomenon in Many Guises." *Review of General Psychology* 2, no. 2 (1998): 175–220.

Nisbett, Richard E., and Timothy D. Wilson. "The Halo Effect: Evidence for Unconscious Alteration of Judgments." *Journal of Personality and Social Psychology* 35, no. 4 (1977): 250–256.

———. "Telling More than We Can Know: Verbal Reports on Mental Processes." *Psychological Review* 84, no. 3 (1977): 231–259.

Norton, Michael I., Daniel Mochon, and Dan Ariely. "The 'IKEA Effect': When Labor Leads to Love." *Journal of Consumer Psychology* 22, no. 3 (July 2012). http://papers.ssrn.com/abstract=1777100.

Peace, Richard. *Conversion in the New Testament: Paul and the Twelve*. Wm. B. Eerdmans Publishing, 1999.

Pearson, Matthew. "Design for Action 2014 Presentation." Design for Action Conference. Washington D.C., October 2014.

Pew Research Center. "When Americans Say They Believe in God, What Do They Mean?," April 25, 2018. https://www.pewforum.org/2018/04/25/when-americans-say-they-believe-in-god-what-do-they-mean/.

Quinn, Jeffrey M., Anthony Pascoe, Wendy Wood, and David T. Neal. "Can't Control Yourself? Monitor Those Bad Habits." *Personality and Social Psychology Bulletin* 36, no. 4 (April 1, 2010): 499–511.

Riet, Jonathan van't, Siet J. Sijtsema, Hans Dagevos, and Gert-Jan De Bruijn. "The Importance of Habits in Eating Behaviour. An Overview and Recommendations for Future Research." *Appetite*, Feeding infants and young children: guidelines, research and practice, 57, no. 3 (December 1, 2011): 585–96.

Rubin, Gretchen. "Strategy of Loophole-Spotting #2: Moral Licensing." Gretchen Rubin, 2014. http://gretchenrubin.com/2014/01/strategy-of-loophole-spotting-2-moral-licensing/

———. *Better Than Before: What I Learned About Making and Breaking Habits—to Sleep More, Quit Sugar, Procrastinate Less, and Generally Build a Happier Life*. New York, NY: Broadway Books, 2015.

Schwartz, Barry. *The Paradox of Choice: Why More Is Less*. New York, NY: Harper Perennial, 2004.

Story, Louise. "Anywhere the Eye Can See, It's Likely to See an Ad." *The New York Times*, January 15, 2007, sec. Media.

https://www.nytimes.com/2007/01/15/business/media/15everywhere.html.

Tan, Jonathan HW. "Behavioral Economics of Religion." *Association of Christian Economists Discussion Papers*, no. 11 (2012). http://nebula.wsimg.com/77a9865c203ca88390cab92fccf430db?AccessKeyId=E117CE44BCF580E0021E&disposition=0&alloworigin=1.

Tangney, June P., Roy F. Baumeister, and Angie Luzio Boone. "High Self-Control Predicts Good Adjustment, Less Pathology, Better Grades, and Interpersonal Success." *Journal of Personality* 72, no. 2 (April 2004): 271–324.

Tesser, Abraham. "The 'What the Hell' Effect." In *Striving and Feeling: Interactions Among Goals, Affect, and Self-Regulation, Tessler Martin and Tesser Eds.*, edited by Winona Cochran. Psychology Press, 1996.

Thaler, Richard H., and Shlomo Benartzi. "Save More Tomorrow™: Using Behavioral Economics to Increase Employee Saving." *Journal of Political Economy* 112, no. S1 (February 2004): S164–87.

Thaler, Richard H., and Cass R. Sunstein. *Nudge: Improving Decisions about Health, Wealth, and Happiness*. New Haven, Connecticut: Yale Univ Press, 2008.

Tversky, Amos, and Daniel Kahneman. "Availability: A Heuristic for Judging Frequency and Probability." *Cognitive Psychology* 5, no. 2 (September 1973): 207–32.

———. "The Framing of Decisions and the Psychology of Choice." *Science* 211, no. 4481 (January 30, 1981): 453–58.

Versland, Amelia, and Harold Rosenberg. "Effect of Brief Imagery Interventions on Craving in College Student Smokers." *Addiction Research & Theory* 15, no. 2 (January 1, 2007): 177–87.

Walker, Ian, Gregory O. Thomas, and Bas Verplanken. "Old Habits Die Hard: Travel Habit Formation and Decay During an Office Relocation." *Environment and Behavior* 47, no. 10 (December 1, 2015): 1089–1106.

Watson, P.C. "On the Failure to Eliminate Hypotheses in a Conceptual Task." *Quarterly Journal of Experimental Psychology* 12 (1960): 129–40.

Wendel, Stephen. *Designing for Behavior Change: Applying Psychology and Behavioral Economics*. Sebastopol, California: O'Reilly Media, 2013.

———. *Improving Employee Benefits: Why Employees Fail to Use Their Benefits and How Behavioral Economics Can Help*. Longfellow Press, 2014.

Wilson, Timothy D. *Strangers to Ourselves: Discovering the Adaptive Unconscious*. Cambridge, Mass: Belknap Press, 2002.

———. *Redirect: The Surprising New Science of Psychological Change*. New York, NY: Little, Brown and Company, 2011.

Wilson, Timothy D, and Daniel T Gilbert. "Affective Forecasting." *Advances in Experimental Social Psychology* 35, no. 35 (2003): 345–411.

Wilson, Timothy D., and Suzanne J. LaFleur. "Knowing What You'll Do: Effects of Analyzing Reasons on Self-Prediction." *Journal of Personality and Social Psychology* 68, no. 1 (1995): 21–35.

Wood, Wendy. "Habit in Personality and Social Psychology." *Personality and Social Psychology Review* 21, no. 4 (November 1, 2017): 389–403.

———. *Good Habits, Bad Habits: The Science of Making Positive Changes That Stick*. New York: Farrar, Straus and Giroux, 2019.

Wood, Wendy, and David T. Neal. "A New Look at Habits and the Habit-Goal Interface." *Psychological Review* 114, no. 4 (2007): 843–63.

———. "The Habitual Consumer." *Journal of Consumer Psychology* 19, no. 4 (October 2009): 579–92.

Wood, Wendy, Jeffrey M. Quinn, and Deborah A. Kashy. "Habits in Everyday Life: Thought, Emotion, and Action." *Journal of Personality and Social Psychology* 83, no. 6 (2002): 1281–97.

Wood, Wendy, and Dennis Rünger. "Psychology of Habit." *Annual Review of Psychology* 67 (2016).

Zajonc, Robert B. "Attitudinal Effects of Mere Exposure." *Journal of Personality and Social Psychology* 9, no. 2, Pt.2 (1968): 1–27.

NOTES AND REFLECTIONS

You can use this space to write about how you want to live your spiritual life, and the obstacles that you're facing on that path. Then, think through creative ways to remove or avoid those obstacles, with the help of spiritual design.

Spiritual Design

Notes and Reflections

Spiritual Design

www.ingramcontent.com/pod-product-compliance
Lightning Source LLC
Chambersburg PA
CBHW070851050426
42453CB00012B/2132